PLASMA SCIENCE
AND TECHNOLOGY

Also by Herman V. Boenig

Unsaturated Polyesters: Structure and Properties
Polyolefins: Structure and Properties
Structure and Properties of Polymers

PLASMA SCIENCE AND TECHNOLOGY

Herman V. Boenig

Cornell University Press

Ithaca and London

First published 1982 by Cornell University Press.
Published in the United Kingdom by Cornell University Press Ltd.,
Ely House, 37 Dover Street, London W1X 4HQ.

Printed in the United States of America

Library of Congress Cataloging in Publication Data

Boenig, Herman V.
  Plasma science and technology.

  Includes bibliographical references and index.
  1. Plasma (Ionized gases) 2. Polymers and polymerization. I. Title.
QC718.B56   530.4′4   81-15200
ISBN 0-8014-1356-7   AACR2

Contents

Preface

This book describes applications of low-temperature plasmas to chemical reactions and, in greater detail, to polymers formed or treated in plasma. Both the mechanisms of polymerization of reactive plasma species and the structure of such polymers are discussed in the light of present knowledge. Separate chapters describe their properties and applications. In addition, modifications of material surfaces by a plasma environment are shown to alter their surface properties markedly either by enhanced crosslinking of polymer surfaces or by introduction into such surfaces of other atoms or groups. Implantation of ions into metal surfaces is also described in detail as a rapidly evolving technique that imparts important properties to metal surfaces.

The unique properties of such plasmas have offered opportunities for processes hitherto inaccessible; plasmas find use as membranes permitting reverse osmosis and have other applications in areas as varied as microcircuitry, integrated optics, etching, detoxification, and thin-layer deposition. Experiments have also demonstrated that in a plasma containing the elements of our earth's atmosphere, for example, Co_2, N_2, and H_2, building blocks for life on earth, such as amino acids, can be synthesized. In addition, we have seen that in an oxygen plasma, organic matter can be completely converted into carbon dioxide and water and that this technique is now considered for use in devices to clean polluted air and to convert waste products in such confined spaces as submarines and space capsules.

Both the complexity and number of chemical reactions and products involving a plasma environment point to the importance of an entirely new field in chemistry, as in physics, the field of plasma chemistry.

I am grateful for the support and encouragement of Lord Kinematics. I also thank my wife, Ilse Boenig, for her help and for many fruitful discussions.

HERMAN V. BOENIG

Erie, Pennsylvania

PLASMA SCIENCE
AND TECHNOLOGY

1. A Brief History

The term "plasma" was first used by Irving Langmuir [1] in 1926 to describe the inner region of an electrical discharge. Later, the definition was broadened to define a state of matter in which a significant number of the atoms and/or molecules are electrically charged or ionized. The first form of plasma observed was the positive column of the glow discharge, in which an equal number of positive ions and electrons are present.

Plasmas differ greatly in many respects, depending on the parameters by which they are generally classified; these include the pressure, charged-particle density, and temperature. Furthermore, the boundary conditions as well as the presence of external electric and/or magnetic fields yield different forms of plasma. The broad definition of plasma as a collection of equal numbers of positive- and negative-charge carriers makes the subject area quite extensive. It does not imply any restriction as to the density of the charged particles, the presence of neutral species, the emission or absorption of electromagnetic radiation, or the motion of particles.

If the positive ions are fixed, as in a solid, and the electrons are mobile, the system may be referred to as solid-state plasma. Liquid plasmas exist in salt solutions in which the positive and negative ions move separately. In this treatise only plasmas in a gaseous state are considered, that is, in a state in which free electric charges can move through the gas, usually under the influence of an electric field, in which the gas is ionized. Langmuir made measurements by inserting small conducting probes into such plasmas. Much later, such "Langmuir probes" were used in rockets and satellites to make the first in situ measurements of ionospheric plasma [2]. In subsequent years such probes have been widely employed to measure ionospheric electron temperatures and densities.

The majority of the universe exists in a plasma state, including the stars, which are almost completely ionized because of their high temperatures. The stars are an example of an equilibrium plasma, in which the ionization is thermally induced and the temperatures of the neutral and charged species are in equilibrium. In laboratory experiments, such an equilibrium plasma is fairly uncommon, since laboratory techniques usually involve nonequilibrium processes, which maintain the ionization by raising some of the charged species to a higher temperature than the neutrals. The most common of these processes is the gas discharge, in which an electrical potential, applied across a gap, provides the selected excitation of the charged species. In the case where most of the ionization occurs by direct electron impact, rather than by thermal channels, the discharge is referred to as glow discharge. The electrical discharge is observed where a gas or vapor becomes electrically conducting. One of the earliest sources of this phenomenon appears to be W. Gilbert's *De magnete, magneticisque corporibus, et de magno magnete tellure*, published in 1600. Gilbert found that a charged conductor loses its charge when brought near a flame and that an electroscope becomes charged when it is connected to a flame. Coulomb [3], in 1785, was probably the first to observe that a charged metal sphere loses the greater part of its charge through the air and not through imperfect insulation. About 1800, Petroff [4] discovered the arc discharge. He observed that when the points of two pieces of charcoal connected to a battery were brought together and then drawn apart, a continuous discharge occurred in the air, forming an arc of light of brilliancy hitherto unknown. The arc persisted even when the air pressure was reduced. In the 1830s Faraday [5] discovered what he called a glow discharge at low gas pressure; the glow discharge consisted of a series of alternate luminous and dark zones that varied in length and color, being sometimes stationary and sometimes in motion in the form of striations. These phenomena occurred in tubes filled with air at a few mm of pressure while the discharge was maintained by a source of potential of about 1000 V. He also observed that the current can pass through a discharge tube filled with a gas at low pressure without exhibiting any luminosity at all. This he referred to as a dark discharge. These phenomena, that is, the dark discharge, the glow discharge, and the arc discharge, can conveniently be considered as the three fundamental types of continuous electric discharge. They are self-sustained, since they can be maintained without the support of an external ionizing agency.

In 1858 Plücker [6] found that a glow discharge at a pressure of 1/100 mm Hg emits cathode rays. The beam colors the gas along its path, and when it impinges on the glass wall of the discharge tube, a green fluorescent spot is produced. Eleven years later, Hittorf [7] observed that these cathode rays can be deflected into a magnetic field; and it was subsequently shown [8] that cathode rays are also deflected by an electric field. Crookes [9] stated in 1879 that "the phenomena in these exhausted tubes reveal to physical science a new world, a world where matter may exist in a fourth state." In 1887 Hertz [10] observed that light emitted by a spark caused an adjacent spark gap to break down more easily. One year later, Hallwachs [11] reported that a zinc plate irradiated with UV light becomes positively charged (as a result, we now know, of the emission of photoelectrons). It soon became evident that the cathode rays must have a mass much smaller than that of the lightest gas atoms, and they were considered particles of negative electricity, for which Stoney [12] proposed the name "electron." By the end of the 19th century it was firmly established that these particles carried a negative charge. In 1886 Goldstein [13] demonstrated the existence of rays of positive ions by passing these "canal rays" into an adjoining chamber through a hole in the cathode of a glow discharge. Subsequent work early in this century supplied additional information about the properties of beams of positive ions, a field of still continuing research.

References

1. H. M. Mott-Smith and I. Langmuir. *Phys. Rev.*, *28*, 727 (1926).
2. G. Hok, N. W. Spencer, and W. G. Dow. *J. Geophys. Res.*, *58*, 235 (1953).
3. C. A. de Coulomb. *Mém. Acad. Sci.* Paris, 612 (1785).
4. W. Petroff. *Galvano-Volta Experiments with a Giant Battery*. Acad. Med. Chir.: Petersburg (1803).
5. M. Faraday. *Researches in Electricity*. London (1844).
6. J. Plücker. *Pogg. Ann.*, *103*, 88 (1858).
7. W. Hittorf. *Pogg. Ann.*, *136*, 1 and 197 (1869).
8. E. Goldstein. *Berl. Ber.*, 271 (1876).
9. W. Crookes. *Phil. Trans.*, *1*, 135 (1879).
10. H. Hertz. *Wied. Ann.*, *31*, 983 (1887).
11. W. Hallwachs. *Wied. Ann.*, *33*, 301 (1888).
12. G. J. Stoney. *Proc. Dublin Soc.*, *4*, 53 (1891).
13. E. Goldstein. *Berl. Ber.*, 691 (1886).

2. The Nature of Plasma

1. General

Plasma can be broadly defined as a system of electric neutrality composed of positive- and negative-charge carriers (see diagram below). The degree of ionization may range from small, as is the case for the reactions discussed in this treatise, to very high, as in the systems studied by physicists in areas such as nuclear reactions. Within these regions, gases may display a wide variety of physical and chemical properties entirely different from those encountered under normal conditions, thus pointing to Crookes's concept, mentioned earlier, of plasma as a fourth state of matter.

```
+ − + − + − + − + −
− + − + − + − + − +
+ − + − + − + − + −
− + − + − + − + − +
+ − + − + − + − + −
− + − + − + − + − +
+ − + − + − + − + −
− + − + − + − + − +
+ − + − + − + − + −
− + − + − + − + − +
```

A number of natural and man-made plasmas were mentioned in chapter 1. One condition for plasma is that it must always be neutral. Therefore, a more specific definition of a plasma is that it is an ensemble of positively and negatively charged particles arranged in such a manner as to shield externally and internally generated electrostatic fields. This electric neutrality is true only in a macroscopic sense. If attention is focused on one single charged particle within a plasma, such as a positive ion, its radial field will induce charge separation in its immediate vicinity. This causes electrons to be attracted toward it and positive ions to be repelled from it. This particle separa-tion creates a potential, V, which is symmetric around the positive ion and is a function of the radial distance from its center.

The concentrations of the positive and negative ions are described by Boltzmann's law of distribution (Eq. 2.1).

$$n^+ = n_p \epsilon^{-eV/kT} \qquad \text{(Eq. 2.1a)}$$

$$n_e = n_p \epsilon^{eV/kT} \qquad \text{(Eq. 2.1b)}$$

where n_p is the density of either charged particle in the plasma as a whole and must be the same for the particles of both polarities to satisfy the requirements for macroscopic neutrality.

Assuming that the potential energy due to charge separation is much smaller than the thermal energy, that is:

$$eV << kT$$

then equations 2.1a and 2.1b become:

$$n^+ = n_p \left(1 - \frac{eV}{kT} \right) \qquad \text{(Eq. 2.2a)}$$

and

$$\lambda n_e = n_p \left(1 + \frac{eV}{kT} \right) \qquad \text{(Eq. 2.2b)}$$

The charged particles within such a system arrange themselves within a distance of a Debye length, λ_D, so that they achieve the local charge neutrality over regions greater than the Debye length. For an electron density N_e at temperature T_e, the Debye length is given by:

$$\lambda = 6.9 \cdot \left(\frac{T_e}{N_e} \right)^{\frac{1}{2}} \qquad \text{(Eq. 2.3)}$$

For a confined plasma the above criteria are satisfied if the physical dimen-sions of the confining system are large compared with the Debye length and if there is a sufficient number of electrons within a sphere with the radius λ_d to produce shielding. For a cylindrical geometry of radius R_0 the plasma criteria are:

$$\lambda_d < R_0$$

and

$$N_e \ll 1.9 \times 10^6 \cdot T_e^3$$

A plasma's behavior is different from that of a neutral gas in that:

Plasmas can be strongly influenced by electric and magnetic fields.
Plasma can conduct electricity, sometimes better than copper.
Long-range forces in a plasma lead to a highly complicated collective behavior.

2. Types of Known Plasmas

Low electron-energy plasmas are sometimes called "cold plasmas." They include interstellar and interplanetary space, the earth's ionosphere, and some alkali-vapor plasmas. In a glow discharge the electron temperature, T_e, is significantly greater than the gas temperature T_g; that is,

$$T_e \approx 1\text{--}10 \text{ eV}$$

at an electron density

$$N_e \approx 10^9\text{--}10^{12}/\text{cm}^{-3}$$

$$T_e/T_g \sim 10\text{--}100$$

In arc discharges (plasma jets), such difference does not exist;

$$T_e \approx T_g \approx 5000°\text{K} - 25{,}000°\text{K}.$$

Figure 2.1 illustrates both the variety and wide range of ionized systems that are categorized under plasmas. It is noted that the plasma regions have defined boundaries between which no plasma is known to exist. In the arc the electron density is much higher than in the glow region. The position of high-pressure arcs in figure 2.1 depends on the gas pressure. In the vicinity of atmospheric pressures, where they are usually produced, the charged-particle density in their core will range from 10^{14} to 10^{18} electrons per cm^{-3}. Because of the considerable number of variables, the characteristics of some plasmas are unknown. Such variables include the nature of the charged particles, their density, their energy, the electric and magnetic fields present, and any energy transfers. Fully ionized plasmas at temperatures of many millions of degrees centigrade have been studied for their importance in cosmology and ther-

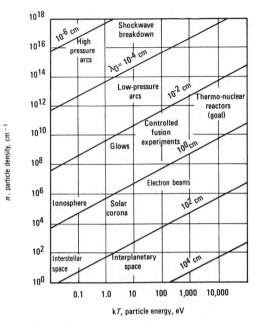

Figure 2.1. Plasma regions. *Source:* Adapted from R. A. Auerbach and D. R. Blenner, Lord Corp., March 23, 1978, and D. T. Tuma, Carnegie-Mellon University, Pittsburgh, Pa., 1979.

monuclear power generations, in which nuclear reactions may contribute to the world's energy supply.

3. Types of Discharges

A. General Description

The following distinctions are frequently made among the silent discharges:

Glow discharges established at reduced pressure by d.c. or low frequency a.c. applied across electrodes sealed into the discharge tube or fixed outside the reactor.

Glow discharges at reduced pressures brought about by the use of higher frequency, such as radio frequencies or microwave frequencies, alternating current passed through a coil surrounding the tube, or between two external electrodes attached to the tube.

Point or point-plate discharges established between points or from a point or points to a plate. Since the electric field is very high at a curved surface of small radius, breakdown of gases can occur in the vicinity of a point at much higher pressures than would normally be the case.

Corona and semicorona discharges occurring through the use of a fine wire instead of points.

Ozonizers as a variation of the corona and widely used for the production of ozone.

B. Ionization

When an atom or molecule absorbs enough energy, one or more electrons are lost, and positively charged particles are formed; the particle is ionized, and the process is called ionization. In contrast, the process of liberating electrons from a solid is generally referred to as electron emission. A gas can exhibit conductivity only when it contains free charged particles, and the degree of conductivity depends on the concentration of those charged particles. The conductivity is a complex function of the number of ions and electrons and their mobility and velocity. In a plasma, where charged particles are normally present together with neutral species, the ionization process must be known if both conductivity and stability are to be evaluated.

Figure 2.2 illustrates several ways by which the various charged particles can be produced by both ionization and electron emission. The most important ionization processes encountered are the ionization of gas particles by electron collision and by absorption of radiation, also known as photoionization. The formation of negative ions, as shown in figure 2.2, occurs when free electrons are available to attach themselves to neutral atoms or molecules. Gases one or two electrons deficient in their outer shells tend to attach an electron, thereby filling the outer shell of the atom and forming a singly charged negative ion. These gases, such as oxygen, are therefore referred to as electronegative gases. Gases that do not have this tendency are known as "free-electron" gases. It is also to be noted that electrons attach not only to atoms but also to molecules of two or more atoms, such as O_2.

When a particle of mass m, that is, an electron, ion, or neutral atom, collides with an atom of mass M, kinetic energy is exchanged. If no excitation or ionization results, such a collision is called an elastic collision. It is inelastic if the gas atom or molecule becomes excited or is ionized by energy acquired from the incident particle. Conversely, a collision between an excited molecule and a neutral particle that yields two particles in a ground state with enhanced kinetic energy is also inelastic. Thus, whenever the potential energy is either increased or decreased, the collision is inelastic. Generally speaking, whenever conversion from kinetic to potential energy occurs, or vice versa, the collision is inelastic. If the kinetic energy of the electron, $\frac{1}{2} mv^2$, is at least equal to the ionization energy of the atom, W_i, that is,

$$\frac{1}{2} mv^2 \geq W_i \qquad \text{(Eq. 2.4)}$$

then ionization should be expected to take place. It has been found, however, that in mercury vapor, for instance, ionization starts not at the ionization

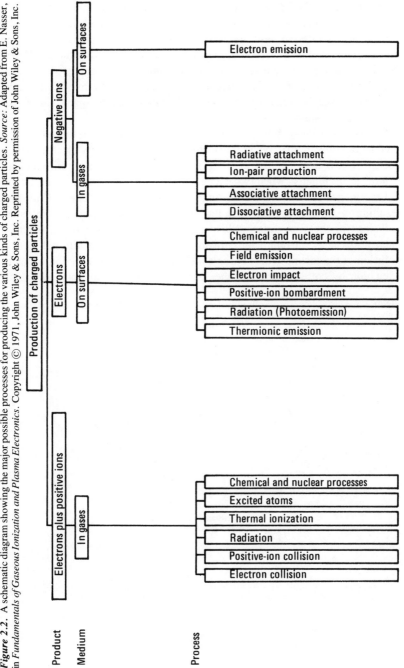

Figure 2.2. A schematic diagram showing the major possible processes for producing the various kinds of charged particles. *Source:* Adapted from E. Nasser, in *Fundamentals of Gaseous Ionization and Plasma Electronics.* Copyright © 1971, John Wiley & Sons, Inc. Reprinted by permission of John Wiley & Sons, Inc.

energy for Hg of 10.38 eV but at an electron energy of 4.77 eV. This energy corresponds exactly to the first excited state of Hg, which happens to be metastable. Before the atom emits two or more photons and returns to the ground state, another electron collides with it, and eventually a third will, and thus ionization may take place. This type of ionization is called "step ionization." It is possible only when the density of the electron beam is high and the atom has metastable excited states whose lifetime is much longer than that of normal excited states.

On the other hand, when an electron collides with an excited atom, the product of collision may be a neutral atom and a high-velocity electron whose energy is now high enough to cause ionization upon collision with another neutral atom. Finally, when two excited atoms collide with each other, potential energy can be exchanged to cause the ionization of one of them. Collisions resulting in ionization caused by the exchange of kinetic energy are called collisions of the first order. When a particle delivers part of its potential energy to cause ionization by collision, this effect is then referred to as a collision of the second order. When a molecule or ion is to cause ionization or excitation by collision with another particle in the ground state, its kinetic

Figure 2.3. Potential effects of collisions. *Source:* Adapted from D. T. Tuma, Carnegie-Mellon University, Pittsburgh, Pa., 1979.

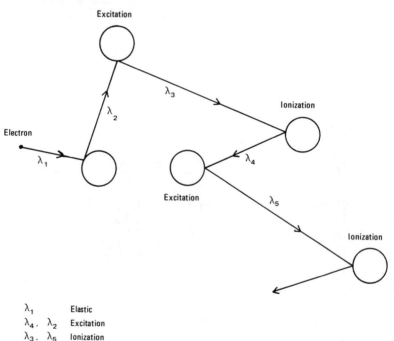

λ_1	Elastic
λ_4, λ_2	Excitation
λ_3, λ_5	Ionization

energy before impact must be at least twice the ionization or excitation energy of the other molecule, provided both particles have the same mass. On the other hand, an electron needs to have a kinetic energy equal only to the ionization or excitation energy of the molecule in order to ionize it by losing all its kinetic energy. Even in such cases all collisions do not result in ionization or excitation. The energy exchange depends on certain probabilities that are usually much lower than unity. In second-order collisions the metastable excited states play a greater role, since their lifetimes, on the order of 10^{-3} to 10^{-2} sec, are longer.

In the analysis of ionization, a quantity called the effective total cross section, $6\ N$, is employed, where 6 is the total target area of one molecule and N is the molecular density, or the number of molecules per unit volume, which is 3.52×10^{22} molecules/m^3 at $t = 0°C$ and $p = 1$ torr. It is found that the mean free path of a particle, $\bar{\lambda}$, is the reciprocal of the effective total cross section;

$$\bar{\lambda} = 1/6N \qquad \text{(Eq. 2.5)}$$

Therefore, $6\ N$ also represents the number of collisions per unit length of path. If the particle undergoes N collisions per unit path, some of them will be elastic, some will result in excitation, and others in ionization (figure 2.3).

C. The Glow Discharge

1. General Description

Figure 2.4 illustrates a qualitative description of the voltage-current relation for a gap, such as that between two plane electrodes of 2 cm diameter with 50 cm separation, at a pressure of 1 torr. When the voltage across the electrodes is increased very slowly, random pulses will be the first measureable current. But when enough free electrons exist in the gap because of external volume ionization, a steady current may be observed with the onset of the pulses.

The current will not be affected by increasing the voltage across the gap until the voltage begins to rise again beyond a certain point. This increase is exponential and is called the Townsend discharge. Further increase of V will result in an overexponential increase in the current, followed by a collapse of the voltage across the gap, which is called "breakdown." It is accompanied by a current increase of several orders of magnitude with almost no increase in voltage. Here the current becomes independent of the external ionizing source and is therefore self-sustained. This is in contrast to regions A, B, and C, where the current becomes zero as soon as the ionizing agent disappears. If the current is increased further by reducing the resistance of the outer circuit,

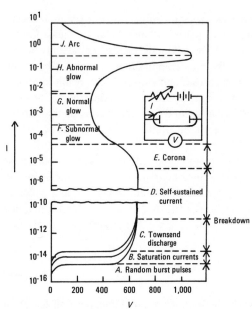

Figure 2.4. Voltage-current relationship in glow discharges. *Source:* Adapted from D. T. Tuma, Carnegie-Mellon University, Pittsburgh, Pa., 1979.

the voltage across the discharge will drop to a low level. Regions F and G are called the "subnormal" and "normal" glow regions, respectively. An exponential increase in current results in the "abnormal" glow.

The glow, also referred to as glow discharge, that readily develops in a gas at pressures of less than 100 torr, is illustrated in figure 2.5 in the various regions between the cathode, on the left, and the anode, on the right. Because of the distinctive features of these regions, they have been given characteristic names. Starting at the cathode, one may observe a thin dark area called the "Aston dark space." It is followed, from cathode to anode, by the cathode glow layers, or sheaths, the negative glow, the Faraday dark space, the positive column, the anode dark space, and the anode glow. The first three constitute the cathode region and the last two the anode region. With the exception of the Aston dark space, the dark spaces are not entirely nonluminous. A small amount of radiation exists in the visible range, but since its intensity is small as compared with the bright regions, the human eye perceives these regions as relatively dark.

Figure 2.5b illustrates schematically the variation of the emitted light in intensity from cathode to anode, showing that the light intensity in the dark regions is not zero and that it varies throughout each region. It is interesting to note that the anode dark space is actually brighter than the cathode glow layer, which appears to be luminous only because it is surrounded by two less luminous regions. The negative glow is the brightest zone of the discharge.

Figure 2.5c shows that the potential V does not vary linearly with the distance from cathode to anode because of the presence of space charges of both polarities; instead, it rises from cathode to anode. Figure 2.5d illustrates that high field exists only near the cathode. Figure 2.5e displays the distribution of positive and negative space charges along the tube. From the cathode to the trailing edge of the negative flow, an excessive positive space charge exists that accounts for the rise in potential, as shown in figure 2.5c. The current densities, j^+ and j^-, illustrate in figure 2.5f that the positive-ion current prevails only in the cathode region, while the electron current is high in all other regions.

2. Effect of Pressure

If the pressure in the tube of figure 2.5, which contains neon at 1 torr, is increased, then the negative glow and two dark spaces that surround it will shrink toward the cathode, while the positive column extends to fill the space. As the pressure is raised further to 100 torr, the positive column continues to

Figure 2.5. Regions in glow disorder. *Source:* Adapted from D. T. Tuma, Carnegie-Mellon University, Pittsburgh, Pa., 1979.

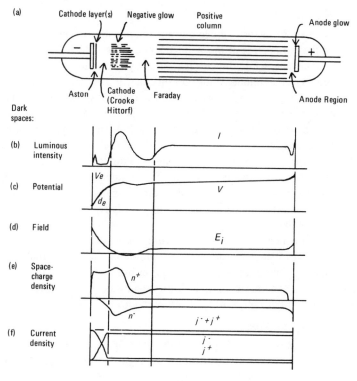

expand, and the region in the cathode vicinity becomes so compressed that the other areas become indistinguishable from one another. Only the negative glow and the Faraday dark space may be recognized with magnifying optics. Furthermore, the positive column has contracted radially, so that it no longer occupies the whole cross section of the tube.

If the pressure inside the tube is reduced below 1 torr, the reverse happens. The cathode regions extend their length at the expense of the positive column, and the boundaries become more diffuse. As the pressure is further reduced, this tendency will continue until the positive column disappears entirely. The glow discharge is generally observed at pressures of 100 torr or less. While it is possible to produce pressures up to atmospheric or higher, continuous cooling of the cathode would be required to prevent the transition from the glow to an arc.

3. Effect of Electrode Distance

As the distance between the electrodes is made larger, a somewhat higher voltage is needed to maintain a glow discharge. The positive column expands to occupy the new volume, and only a small change in the dimensions of the other regions takes place.

4. Effect of Voltage

It can be seen from the voltage-current relationship (figure 2.4) that by variation of the parameters of the voltage source, the voltage across the tube will remain constant, while the current varies by two, and in some instances by three, orders of magnitude. If the current exceeds a certain value, however, the voltage will increase. At a low current the cathode glow layer extends to only a part of the cathode surface. If the current is reduced further, the voltage must be raised to maintain the glow of low light intensity. This type of discharge is referred to as the subnormal glow. If the current is increased by reducing the series resistance of the outer circuit, the cathode glow area extends proportionally with the current. This tendency continues until the whole cathode surface is covered by the cathode glow. Therefore, both the current density and the voltage across the tube are fairly constant. This region is referred to as the "normal glow."

If the current is increased further by raising the voltage across the tube, the cathode glow will establish itself on any additional surface available. This state is called the "abnormal glow"; it requires higher voltage to maintain the current, as indicated in figure 2.4. The brightness of the luminous parts increases with the current in all these glow modes.

5. Effect of the Nature of Gas

Although the previous description of the glow at 1 torr pressure applies to neon, the general characteristics of the discharge do not change very much with the type of gas. The most dramatic change is in the color of the three

Table 2.1. Characteristic colors of the glow regions in some gases

Gas	First cathode layer	Negative glow	Positive column
Air	pink	blue	—
H_2	brownish red	pale blue	pink
N_2	pink	blue	red
O_2	red	yellowish white	pale yellow with pink center
He	red	green	red to violet
A	pink	dark blue	dark red
Ne	yellow	orange	brick red

Source: E. Nasser, *Fundamentals of Gaseous Ionization and Plasma Electronics,* Wiley: New York (1971). Copyright © 1971, John Wiley & Sons, Inc. Reprinted by permission of John Wiley & Sons, Inc.

bright zones. In the above example, the colors of the cathode layer, negative glow, and positive column are yellow, orange, and brick red, respectively. As is shown in table 2.1, these colors are distinctive characteristics of the gases [1]. When the cathode has a low work function, the voltage needs to be reduced to maintain constant discharge.

6. The Cathode Zone

This zone of the discharge carries most of the voltage and is, therefore, of considerable importance. Its boundaries—the cathode surface itself and the bright edge of the negative glow—are quite distinct. The distance between these boundaries, d_c, and the voltage across it, V_e, is generally referred to as the "cathode fall." The physical processes that take place in the cathode zone are nearly independent of all other regions. A glow discharge can exist without a positive column, Faraday dark space, or even negative glow. It can never exist without a cathode dark space, however.

It should be noted that glows exist that do not exhibit a cathode dark space. This is because of the different conditions that sometimes occur at the cathode, when a sufficient supply of fast electrons is provided that cannot be accelerated to the point where they are able to cause ionization. This may occur in cases where a heated cathode provides adequate thermionic emissions or where a cathode with a thin insulating film is used to allow for the accumulation of positive ions and the acceleration of the electrons through the thin film.

The physical process can be described as follows. Electrons are emitted from the cathode by two mechanisms: the bombardment of positive ions, γ_i process, and incident radiation, γ_p process. Because their initial energy is low, the electrons form a thin sheath of electron space charges in which the current is carried mainly by the positive ions moving toward the cathode. As the electrons leave this thin sheath, they are accelerated in the high field in a region of net positive charge.

Electrons and/or negative ions formed in the Aston dark space may be recombined outside this region, inducing a release of ionizative energy in the form of a photon and a layer of luminosity, the cathode glow layer. As the electrons gain energy, negative ions may release their electrons and thus reduce the probability of recombinations in the high-field region of negative glow. The acceleration depends on the magnitude and direction of the field, which is axial in this case. With a cathode of different geometry, both electron motion and energy gain will be different, and glow discharges of "unusual" properties may be observed. As an example, a hollow cylinder serving as a cathode, called a "hollow cathode," produces extreme brightness in the negative glow zone and is characterized by a high current that is orders of magnitude higher than that of a plane cathode for the same cathode fall [2]. Glows from such hollow cathodes are suitably used as light sources and in other applications [3,4].

7. Negative Glow and Faraday Dark Space

The negative glow is the brightest part of the glow discharge. Its properties are related to the Faraday relatively dark space. The brightness is caused by the energetic electrons arriving at the end of the cathode dark space where they have gained their maximum energy. At the boundary of the negative glow they incur inelastic collisions, and because of their high energies, they induce excitation and ionization. As the electrons are slowed down, both excitation and ionization rates decrease and thus reduce the brightness of the negative glow until it merges into the Faraday dark space. While most ionization occurs by electron collision, photoionization enhances the extent of ionization in the negative glow [5].

The near-absence of emission in the Faraday dark space is ascribed to the low energy of the electrons when they leave the negative glow, so that few or no ions are generated when they emerge from the negative glow. Therefore, the Faraday dark space will generally exhibit an excessive negative space charge. Thus the electrons, after having expended their energy, will not be able to acquire new energy in this region.

8. The Positive Column

The positive column is used in many applications, including light sources, plasma sources, and plasma torches. It represents the luminous part of the discharge between the anode dark space and the edge of the Faraday dark space (figure 2.5). Its color is characteristic of the gas (table 2.1) and varies slightly with the field. It appears to exhibit a steady and uniform luminosity, although it may contain fast-moving striations in both directions. The gas in the positive column is in an ionized state with equal densities of positive and negative particles. This is the condition generally referred to as "plasma"; a plasma is an ionized gas in which there are equal numbers of positive and negative particles.

9. The Anode Region

The anode in the glow discharge does not emit any particles; it merely collects the electrons. Normally, a negative space charge exists in the vicinity of the anode, which is relatively dark. A voltage drop across this zone, called the "anode fall," is on the order of the ionization potential of the gas. In this region, the ions are generated that enter the positive column. This is accomplished by the electrons emerging from the positive column and gaining energy in the anode fall, thus causing a light glow at the anode surface, called the anode glow (figure 2.5).

10. Effect of Electromagnetic Radiation

Generally, chemical reactions require the input or transfer of energy. One important technique of transforming such energy is the employment of electromagnetic radiation. Transfer of energy by radiation can be a useful technique for creating those chemical reactions that require avoidance of contamination that may take place via contact with a heating medium or electrode.

Electromagnetic radiation spans a large spectrum of frequencies, such as those found in the order of increasing frequency in radio waves, infrared, visible light, and ultraviolet radiation. Energy sources utilizing lower frequency ranges have been reported extensively. Frequencies of the energy sources that induce chemical reactions have included electrical energy, thermal energy, and electromagnetic energy [6]. Other investigators [7] describe the use of induction plasmas with a frequency range from 50 Kc to 15 Mc for conducting chemical syntheses. The range of microwave frequencies from about 10^3 Mc to 10^6 Mc has been studied during the last 35 years.

Radiowave and microwave discharges have become particularly important in investigations involving electrically induced chemical reactions, since this technique permits chemical reactions with the presence in the discharge zone of electrodes. Figure 2.6 illustrates several techniques for coupling high-frequency power into a discharge.

Both inductive and capacitive systems are presently employed for radio-frequency discharges. At microwave frequencies, sufficient coupling is obtained by passing the discharge tube directly through a section of wave guide. In such high-frequency discharges, energy is transferred from the electric field by electrons generated by the ionization of neutral gas molecules or atoms. These electrons gain energy from the field while they are involved in collisions with gas molecules. The collisions tend to convert the oscillating motion of the electrons into a random motion. In order to restore the ordered oscillatory motion, the field induces a network on the electrons and in this way

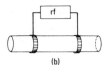
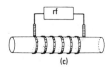

Figure 2.6. Typical arrangements for electrodeless high-frequency discharges: (*a*) plates outside the reactor, (*b*) rings, and (*c*) a coil around the reactor. *Source:* Adapted from R. A. Auerbach, Lord Corp., 1980.

enhances their energy. The averages in power, P, transferred per unit volume gas through this mechanism is

$$P = \frac{n_e^2 E_o^2}{2m} \cdot \frac{v_c}{v_c^2 + \omega^2} \qquad \text{(Eq. 2.6)}$$

where E_o is the amplitude of the h-f field strength, m is the mass of the electron, n is the electron density, v_c is the elastic collision frequency, and ω is the frequency of the applied field. The magnitude of v_c lies between 10^9 and 10^{11} collisions per sec. Thus, for r.f. frequencies where $\omega \approx 10^7$ and $v_c \gg \omega$, the effect of the driving frequency in equation 2.6 disappears. The electrons generated by ionization of gas may disappear from the discharge zone by diffusion to the walls of the container. A steady state is maintained when the rates of production and loss balance.

For an r.f.-sustained hydrogen discharge at 10 torr in a tube 1 cm in diameter, the following conditions are obtained [8]:

$E = 125$ volts/cm, $n = 3 \times 10^{10}$ electrons/cm³, and the current density, j (which is proportional to the product of the electron density, n, and the field strength, E) = 10 mA/cm², $\bar{P} = 1.25$ watts/cm³.

These values are similar to those of the positive column of the d.c. discharge points. It has been shown experimentally [9] that the elementary theory of the positive column of d.c. discharges can be employed to predict the plasma parameters of an r.f. discharge. In fact, Golan [10] has demonstrated earlier that the kinetic equations for the d.c. and high-frequency discharges become identical if

$$\omega \gg \frac{m}{M} v_c$$

$$E_{dc} \approx \frac{E_{oac}}{\sqrt{2}} \cdot \frac{v_c^2}{v_c^2 + \omega^2} \qquad \text{(Eq. 2.7)}$$

and if the applied fields are perpendicular to the electron-density gradients as well as to the space-charge fields.

11. Other Phenomena

In addition to the normal phenomena discussed in the preceding paragraphs, there are processes that take place under special conditions. They include cathode sputtering, cataphoresis, obstructed glows, and positive column striations. These phenomena are beyond the scope of this treatise and are described elsewhere [1].

D. The Corona Discharge

A crownlike form of a glow in points or along wires in air raised to high potentials has led to the name of this type of discharge. It occurs at any pressure in all gases but is particularly pronounced if the pressure is relatively high. The corona discharge can be observed on lightning conductors or ships' masts when densely charged clouds are discharging in close vicinity and can also appear on high-voltage transmission lines. It is widely used to produce electric charges in gases and in ozonizers for sterilizing water. This phenomenon is physically similar to a glow discharge in a highly nonuniform electric field.

The characteristic distinguishing a direct-current discharge from the glow discharge is that the corona discharges a glow without positive or negative zones. The electric energy in a corona is converted chiefly into heat in the gas; the ions impart their momentum to molecules. Only a small fraction of the energy is converted into chemical energy, and light losses by recombination are believed to be small.

The alternating-current corona is complicated by the fact that the space charge is driven into the space by the field of the wire. When the field changes its direction, this space charge is attracted by the wire, but at the same time ions of opposite polarity are repelled. Therefore, in the steady state, there exists a marked phase difference between the field of the surface of the wire and the velocity of the ions. This results in a corresponding phase difference between the peak value of the field and the corona current output. The alternating-current corona is of practical interest, since the losses of power on transmission lines caused by such discharges can be considerable.

E. Arc Discharge

When two carbon or metal electrodes are brought into contact and then separated in a circuit of sufficient ampere flow, a self-sustained arc discharge takes place. This type of discharge occurs in atmospheric air as well as at lower or higher pressures and in various gases and vapors.

In a typical voltage-current characteristic of a gas, such as that illustrated in figure 2.4, the last region of a discharge, that is, the region having the highest current, is the arc (in region J of figure 2.4). In this region the current is about two orders of magnitude higher than in the glow, while the voltage is approximately one order of magnitude lower. Although the arc has been used for a long time, it is considered to be the form of discharge least understood theoretically. An arc can be defined by comparing its cathode region with that of a glow discharge. Here, the cathode has a potential fall of about 100–400 volts and a low current density. Except at high pressures and currents, the positive column always fills the tube. Thermal effects do not contribute to the functioning of the cathode of a glow discharge, and the light emitted from the region near the cathode has the spectrum of the gas.

On the other hand, the arc cathode has a fall of potential in the order of 10 V, a very high current density (a demand for thermal effects essential for its working), and light emission that has the spectrum of the vapor of the cathode material. The positive column is constricted near the electrodes. The glow-to-arc transition occurs when the abnormal glow has acquired a critical voltage value and when the power supply has sufficiently low internal resistance capable of sustaining the high currents needed for the arc. It can be observed from figure 2.4 that the slope of the voltage-current relation of the arc is negative: an increasing current results in a drop in the voltage across the arc.

Arcs have been known since the early days of electricity, and their high light intensity is still being used for lighting purposes. Because of the arc's high temperature, extensive research is being conducted to produce the temperature needed to trigger thermonuclear fusion. Arcs are produced in the vicinity of atmospheric pressures where the charged-particle density in their cores ranges between 10^{14} and 10^{18} electrons/cm^3 (figure 2.1). The temperatures in the arc range from 200°K at 10^{-3} torr to 4000°K at atmospheric pressure.

4. Plasma Temperature

Kinetic theory established early a relationship between the mean translational energy of the molecules in a gas and temperature; that is, temperature can be expressed in terms of energy. Molecules exist in other forms of energy, such as rotational and vibrational energy, however. Furthermore, molecules in a gas do not all possess the same velocity; rather, they follow the Maxwellian distribution of velocities, and it has been observed that such a form of distribution also applies to plasma particles, provided that electrical currents of high-frequency fields are absent.

Plasma, by definition, consists of ionized gases—positively and negatively charged particles. The latter, being generally electrons and therefore much smaller in size than the positive ions or neutral species, will have a much greater velocity, and therefore a much higher temperature, than ions and molecules. The ions will also exhibit higher velocities, and therefore temperatures, than the corresponding neutral species. As a consequence, the species in a plasma may vary widely in temperatures. In addition, there is no technique available that permits a direct temperature measurement of the various species existing in plasma. Nevertheless, some methods, such as the Doppler broadening of spectrum lines, permit the measurement of the velocities of atoms and molecules. Thus, one characteristic feature of glow discharges is that the Boltzmann temperature of the ions and molecules is roughly ambient in most cases, while that of the electrons is some two orders of magnitude greater. Therefore, a plasma may be characterized in terms of the average electron temperature and the charge density within the system. For simple systems, such as inert-gas plasmas, the solution of the relevant equations has been shown to lead to a Maxwellian distribution (figure 2.7).

Figure 2.7. Maxwellian distribution. *Source:* H. V. Boenig, Lord Corp., 1979.

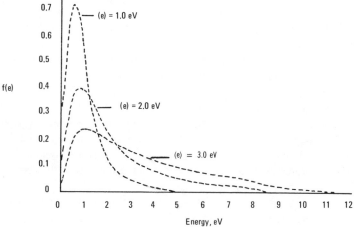

The electron temperature for a Maxwellian distribution can be defined as:

$$\epsilon = 3/2KT_e = \tfrac{1}{2}m\bar{V}_r^2 \qquad \text{(Eq. 2.8)}$$

where K is the Boltzmann constant and $\bar{V}_r$ is the random velocity. The mean energy $T_eV = 7733°K$. Electron temperatures in discharges vary greatly from about 15000°K in mercury vapor rectifiers to about 25000°K in neon tubes and to more than 100,000°K in other plasmas. Thus, the electron-volt temperature is used as a more meaningful measurement than "plasma temperature." For "real systems," which are considerably more complex, the form

Table 2.2. Energies in a glow discharge versus typical bond energies

Energy	eV
In glow discharge	
Electrons	0–20
Ions	0–2
Metastables	0–20
UV/visible	3–40
In bond	
C—H	4.3
C—N	2.9
C—Cl	3.4
C—F	4.4
C=O	8.0
C—C	3.4
C=C	6.1

Source: A. V. Engel, *Ionized Gases*, Oxford University Press: Oxford (1955).

of the distribution has been analyzed by probe measurements [11] and direct electron sampling [12].

Plasmas are copious sources of the electromagnetic radiation, particularly in the UV and vacuum UV, and are indeed frequently used as sources in those regions. Also, the relatively small output in the visible region gives rise to characteristic colors for plasmas excited in a given system, hence the appellation "glow discharge." The reactive species in a plasma resulting from the ionization, fragmentation, and excitation processes that arise from collisions involving electrons and the electric field constitute a relatively complex system. The energies available in a glow discharge together with some typical bond energies are given in table 2.2.

In chemical reactions it is also frequently helpful to relate electron volts to bond energies in terms of Kcal/mole, that is, 1 eV = 23.1 Kcal/mole.

References

1. E. Nasser. *Fundamentals of Ionization*. Wiley: New York (1971).
2. A. V. Engel. *Ionized Gases*. Oxford University Press: Oxford (1955).
3. V. S. Borodin and I. M. Kagan. *Soc. J. Tech. Phys.*, *11*, 131 (1966).
4. D. J. Sturges and H. J. Oskam. *J. Appl. Phys.*, *37*, 2405 (1966).
5. R. F. Pottie, M. J. Vasile, and P. Lightman. *Bull. Amer. Phys. Soc.*, *II-14*, 256 (1969).
6. E. W. R. Steacie. *Atomic and Free Radical Reactions*, Vol. 1. Reinhold: New York, 13–70 (1954).
7. T. B. Reed. *Chemical Uses of Induction Plasmas*. Lincoln Laboratory, MIT: Cambridge, Mass. (1964).

8. A. Bell. In *Application of Plasmas to Chemical Processing*, ed. R. F. Baddour and R. S. Timmins. MIT Press: Cambridge, Mass. (1967).

9. J. T. Massey, A. G. Schulz, S. M. Cannon, and B. F. Hochheimer. *J. App. Phys., 36,* 1790 (1965).

10. V. E. Golan. "Izvest. Akad. Nank." (USSR). *Ser. Fiz., 23,* 952 (1959).

11. E. D. Johnson and L. Malter. *Phys. Rev., 80*(7), 58 (1950).

12. D. T. Clark, H. R. Thomas, and A. Dilks. Paper presented at the Centennial Meeting of the Amer. Chem. Soc., New York, N.Y., April 1976. Amer. Chem. Soc.: Washington, D.C.

3. Glow-Discharge Chemistry

1. Techniques

To the chemist, a plasma provides a new way of transferring energy to molecules. For many reactions it does not matter whether the energy necessary for a given reaction is transferred through ion collision, light absorption, or electron impact. Neither is the type of plasma reactor of significance in most cases, provided that equal electron energies and gas temperatures are obtained. It also appears at this time that there are no discharge types or frequencies that favor specific reactions. Therefore, the criteria for choosing plasma equipment are based more on practical considerations such as constancy in performance, ease of handling, simplicity, size, and cost.

Various techniques have been employed to produce a plasma suitable for chemical reactions, such as silent discharges, direct-current and low-frequency glow discharges, and radio-frequency and microwave discharges. Silent discharges, as typified by those observed in an ozonizer, which require high-voltage power (> 1000 KV) but sustain a silent discharge at line frequencies of 50–60 Hz, have the disadvantage of requiring small gaps and large surface areas. Organic reactants may accumulate on the reactor walls and can cause problems in cleaning.

Direct current and low frequency in the high-voltage glow range require two metal electrodes inside a reactor and operate generally at a voltage of 10–100 V/cm at about 0.1–1 torr and power levels up to 50 W. The disadvantage of this technique is that in organic reactions, the electrodes are readily covered with the products of reactions, such as polymer films. This contamination can be prevented, however, by isolating the electrodes using a shroud of rare gas [1]. At frequencies above 1 MHz, direct contact between the electrodes and the plasma is no longer needed. The energy can be supplied to the plasma indirectly by capacitive or inductive coupling.

The electrodes enclose the plasma tube for inductive coupling, and the tube lies on the axis of the coil. Since the coupling components are situated outside

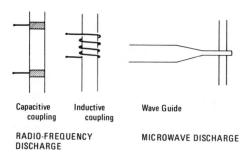

Capacitive   Inductive   Wave Guide
coupling   coupling

RADIO-FREQUENCY   MICROWAVE DISCHARGE
DISCHARGE

Figure 3.1. Typical arrangement for electrodeless discharges. *Source:* Adapted from R. A. Auerbach, Lord Corp., 1980.

the reactor, they are not subject to contamination. Reactor designs of the type shown in figure 3.1 are generally limited to frequencies below 200 MHz. Since the frequency dependence of plasma reactions is of little importance, the choice of operating frequency is decided mainly by cost and aspects of handling and shielding. In the United States the commercially available plasma equipment operates at the frequency of 12.56 MHz, as released by the FCC.

In the past, microwave discharges have chiefly been employed with inorganic compounds [2], while organic compounds are also used in more recent times. Since the dimensions of such a reactor are on the order of the field wavelength, the magnitude of the electrical field can vary significantly. Thus, in some regions the starting material may be completely destroyed, while in others the discharge can hardly be sustained. The decomposition of organic material results frequently in carbon deposits on the walls, which in turn absorb a considerable fraction of the incoming energy. Furthermore, microwave discharges are difficult to maintain below pressures of 1 torr [3]. More recently, however, improved techniques and controls have made it possible to obtain reproducible chemical reactions, including polymer depositions.

The different kinds of electrical discharges have generally been found to lead to similar results. For organic chemical reactions, radio frequency is used most frequently in both laboratories and industrial production, chiefly because of the convenience of handling, ease of availability, and low energy requirements.

2. Plasma Components

Conditions in glow-discharge plasmas as used in both laboratories and industrial applications are most frequently as follows:

Electron temperature	≈ 1–10 eV
Electron density, n_e	$\approx 10^9$–10^{12}/cm^{-3}
Power output	5–1000 watts

Operating radio frequency	13.56 MHz
Pressure conditions	0.1 to 1.0 torr
Suitable gas flow	~ 10 cc/min

The temperature in such a plasma, or the energy of the species, varies with the distance of the electrodes, the power in watts, the gas pressure, and the type of gas and/or vapor.

In all glow discharges the electrons, generated largely by partial ionization of the molecules in plasma, are the principal sources for transferring energy from the electric field to the gas. Elastic electron-molecule collisions will cause an increase in the kinetic energy of the molecules, while inelastic collisions generally result in excitation, fragmentation, or ionization of the molecules (figure 2.3). The plasma components in a reaction tube charged with molecules as derived from the energy, free electrons, and collisions with molecules are illustrated in figure 3.2.

Figure 3.2. Plasma components. *Source:* Adapted from D. T. Tuma, Carnegie-Mellon University, Pittsburgh, Pa., 1979, and H. V. Boenig, Lord Kinematics, 1979.

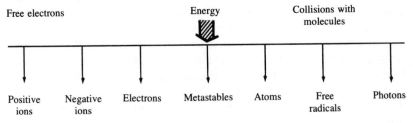

3. Physical Processes

Glow discharges are generally employed in the pressure range of about 0.1 to 10 torr. At much lower pressures, the electron mean free path is too long for gas collisions to be significant. At much higher pressures, the mean free path is very short, and the breakdown field strength is very high, and can generate local and highly ionized but narrow pathways for the conduction of current, and thus can form spark filaments.

The normal reactions occurring in d.c. glow discharges are complex and have been described in chapter 2. In high-frequency electrodeless discharges, the complications of the cathode and anode regions are absent, and the entire plasma is approximately neutral and diffusion-controlled. The discharge is often similar to that observed in the positive column of a direct-current discharge. There are some fundamental differences, however, especially at microwave frequencies.

Free electrons oscillating in an alternating field cannot derive power from the field, since their motion is 90° out of phase with the field. Therefore, they acquire energy only because collisions with neutral molecules alter their phase relationship with the field while simultaneously representing a small fractional loss of the energy gained. Assuming that the a.c. frequency, f, is greater than the elastic collision frequency, ν_c, the maximum displacement, x, of an electron due to the high-frequency field is given by:

$$ x = \frac{2eE}{m\omega^2} \qquad \text{(Eq. 3.1)} $$

where $\omega = 2\pi f$.

In glow discharges as they are discussed in this treatise, that is, where the concentration of charged species is greater than 10^8 cm^{-3}, a large fractional separation of electrons and positive ions becomes impossible, since it would create a large opposing field. The currents of electrons and ions reaching the wall must then be equal (equations 3.2a and 3.2b):

$$ i_e = -D_n \nabla n - n\mu_e E', \qquad \text{(Eq. 3.2a)} $$

$$ i_+ = -D_+ \nabla n + n\,\mu_+ E' \qquad \text{(Eq. 3.2b)} $$

where $n = n_e = n_+$ is the electron density, ∇n is the density gradient, D is the diffusion coefficient, μ the mobility, and E' the field caused by the small space charge. The subscripts e and + represent electron and positive ions. With $i_e = i_+$ and by eliminating E', equations 3.2a and 3.2b become

$$ i = -\frac{D_e\mu_+ + D_+\mu_e}{\mu_+ + \mu_e} \cdot n = D_a \nabla n \qquad \text{(Eq. 3.3)} $$

where D_a is the ambipolar diffusion coefficient. By substituting $\mu_e \cdot kT_e/e$ for D_e and the equivalent expression for D_i, one obtains equation 3.4:

$$ D_a = \frac{\mu_e \cdot \mu_+}{\mu_e + \mu_+} \cdot \frac{k}{e} (T_e + T_i) \qquad \text{(Eq. 3.4)} $$

which approximately equals $\mu_+ k/e\,(T_e + T_i)$ or $D_+ (1 + T_e/T_+)$ because the electron mobility μ_e is considerably larger than the mobility of ion, μ_i. In glow discharges T_e/T_+ is characteristically on the order of 20 to 100, while in the afterglow (see below) the electrons thermilize rapidly, T_e/T_+ becomes unity, and D_a, which is $\approx$ 20 to 100 D_+ in the active discharges, becomes $2D_+$ in the afterglow.

This disappearance of charged species by ambipolar diffusion in the absence of a source term is defined by the diffusion equation 3.5:

$$\frac{\delta n}{\delta t} = D_a \nabla^2 n \qquad \text{(Eq. 3.5)}$$

In a glow discharge several radiative, two-body, and three-body recombination mechanisms are recognized. Those occurring at greatest speed are the dissociative recombination of electrons and positive molecular ions such as illustrated by

$$NO^+ + e \rightarrow N + O$$

whereby the products may be electronically excited.

Taking hydrogen as an example and denoting p and q the principal quantum numbers of the state of the hydrogen atom, a number of recombinations can take place, such as:

$$H^+ + e + e \rightarrow H(p) + e \qquad (1)$$

and the inverse collisional ionization:

$$H(p) + e \rightarrow H^+ + e + e \qquad (2)$$

superelastic collisions:

$$H(p) + e \rightarrow H(q) + e \quad (q < p) \qquad (3)$$

inelastic collisions:

$$H(p) + e \rightarrow H(q) + e \quad (q > p) \qquad (4)$$

downward cascading:

$$H(p) \rightarrow H(q) + h\nu \qquad (5)$$

upward transitions by line absorption:

$$H(q) + h\nu \rightarrow H(p) \qquad (6)$$

radiative recombination:

$$H^+ + e \rightarrow H(p) + h\nu \qquad (7)$$

and photoionization:

$$H(p) + h\nu \rightarrow H^+ + e \tag{8}$$

Furthermore, ion-ion recombination reactions such as $I_2^+ + I \rightarrow I_2 + I$ or $3I$, $NO^+ + NO_2^- \rightarrow$ neutral products, as well as three-body ion-ion recombinations such as $NO^+ + NO_2 + M \rightarrow$ neutral products, are encountered in a glow discharge.

Due to the long-range Coulombic attraction between reactants, these processes have very large rate constants, provided that reasonable paths are available for the dissipation of the large exothermic reaction energy ($\sim 10\,eV$) such as dissociation and electronic excitation. For instance, for the rate constant of

$$N_2 + e \rightarrow N + N + e$$

a value of 2.9×10^{-7} cc molecule^{-1} sec^{-1} was found [4,5]. The general range of 1 to 5×10^{-7} means that at an electron and ion concentration of 10^{11} cm^{-3}, the effective first-order rate constant for electron removal under steady-state conditions will be 1 to 5×10^4 sec^{-1}.

In electron attachment and detachment processes, radiative attachment and photodetachment as well as three-body attachment processes are not considered to be important. Dissociative attachment reactions such as $e + O_2 \rightarrow O^- + O$ show rate constants that increase from zero at an electron energy threshold (of 4 to 9 eV for the formation of O^- from O_2, NO, or CO) to a maximum of 10^{-11} to 10^{-10} cc $\cdot$ molecule^{-1} sec^{-1} for electrons with 6 to 10 eV.

Charge transfer and ion-molecule reactions have small or no activation energy if they are exothermic. This is due to strong ion-dipole and ion-induced dipole interactions. Their rate constants range frequently near 10^{-9} cc molecule^{-1} sec^{-1}, and therefore the effective first-order rate constant for transforming an ionic species by reaction with a neutral particle is 2×10^6 to 2×10^7 sec^{-1} (at $p = 1$ torr $\cong 2 \times 10^{16}$ molecules cm^{-3} at the higher discharge temperature); such reactions will go to completion in a small fraction of the residence time.

In electron-impact ionizations the electron loss term by ambipolar diffusion has been estimated by a first-order rate constant, $k = 5.78 D_a / r_o^2$. For an active discharge with $T_e/T_+ \sim 50$, $D_+ \sim 100$ cm^2 sec^{-1}, and $r_0 = 0.5$ cm, k becomes approximately 1×10^5 sec^{-1}. It is generally accepted that electron-impact ionization is the major source term for charged species in the discharge. Assuming Maxwellian distribution, the total effective ionization rate constants are on the order of 10^{-11} to 10^{-10} cc molecule^{-1} sec^{-1} for average electron energies of 2 to 3 eV. In mechanisms involving direct ionization by collision with energetic metastable neutral species, also referred to as Penning ionization, the ionization energy of the metastable must exceed the ionization potential of the other reactant. For He, Ne, and Ar, the excitation energies of the lowest triplet state are 19.80, 16.62, and 11.55 eV, respectively. Since the rate constants for Penning reactions are frequently gas kinetic, that is, on the order of 10^{-10} cc molecules sec^{-1}, this mechanism is probably the major ionization source in diatomic gases mixed with an excess of He or Ne but not Ar.

In excitation and dissociation processes, rotational excitations are neglected because rotational energies are small. The collisional exchange of translational and rotational energy is so rapid that rotational states are not considered to be significantly out of kinetic temperature equilibrium. Vibration-translation interchange can result in appreciable vibrational excitation and effective vibrational temperatures ranging between T_e and T_g, however. Electron-impact excitation of vibrational energy is a very specific mechanism the cross section of which may vary by two orders of magnitude from one molecule to another, since it is dependent on the existence of negative ion states that decay into vibrationally excited ground-state molecules and slow electrons. This process is especially significant in molecules that have large cross sections on the order of 1 to 5×10^{-16} cm²/molecule ($k = 5 \times 10^{-9}$ to 2×10^{-8} cc molecule^{-1} sec^{-1} at electron energies of about 2 to 5 eV), such as N_2, CO, and CO_2. It is less important in smaller molecules such as H_2 and O_2, whose cross sections are smaller by a factor of 10 to 50 than those in N_2. The high vibrational temperature, T_v, of N_2 is believed to be responsible for a large portion of its electronic excitation, because if T_v approaches T_e, considerable concentrations of molecules having 5 to 8 eV of vibrational energy are likely to be generated.

Electronically excited states can arise directly from electron impact. When the corresponding radiative transmission is allowed, the excitation cross section rises to a broad maximum and then slowly decreases. Peak cross sections are frequently on the order of 10^{-17} to 10^{-18} cm²/molecule. This corresponds approximately to rate constants of 10^{-11} to 10^{-10} cc molecule^{-1} sec^{-1} for transitions requiring 5 to 10 eV, if the average electron energies are 2 to 3 eV.

The important electron production processes take place in the gas phase (and at electrode surfaces if they are present) by simple electron-impact ionization:

$$e + A \rightarrow 2e + A^+$$

where A is a neutral molecule or atom in the ground state. Since only electrons in the high-energy tail of the electron energy distribution have sufficient energy to ionize, and the first few eV above threshold ionization cross sections are nearly linear with energy, the rate coefficient for ionization can be calculated [6].

Such high-temperature electrons can be even more effective in generating excitation than in producing ionization. The excitation of an electron impact is chiefly responsible for the light emitted from the discharge and, at the same time, for the generation of active neutral species.

The result from simple excitation

$$e + A \rightarrow e + A*$$

(9)

depends on the lifetime of $A*$. For radiating states the lifetime is of the order of 10^{-8} sec, which is too short for anything to happen before a photon is emitted and the atom has returned to a lower state. In most cases the light will simply escape. If the radiation occurs to the ground state, however, then the photon can be absorbed by another atom or molecule in the vicinity, which is then excited. This process, called resonance radiation, is essentially equivalent to having an excited atom the lifetime of which is equal to the time needed for the photon to escape the discharge, which can be considerably longer than the natural lifetime of the excited state.

Excitation can also result in the formation of true metastable states, whose lifetime against radioactive decay is long, that is, μsec to sec. Such metastable neutrals will subsequently be destroyed by collision processes in the plasma.

Neutrals with high electronic excitation can readily be ionized by both electrons and photons. They can also ionize neutrals whose ionization potential is smaller than the excitation energy. This process is often referred to as "Penning ionization."

From the standpoint of chemistry, it is interesting to speculate on reactions initiated by excited neutrals, which are forbidden to ground-state neutrals.

If a molecule is to be excited, dissociation decay could be considered:

$$e + AB \rightarrow e + (AB)* \rightarrow e + A + B$$

(10)

In this event, the electron energy must exceed the dissociation energy. This is an effective mechanism for the generation of free radicals. It would seem, therefore, that electron-impact excitation is of major importance to gas-discharge chemistry. Except in a qualitative way, such excitation processes are far from fully understood.

In discharge chemistry, major wall effects deal not with electron production and loss mechanisms but with the surface catalytic effects of walls. Little is known about such processes, however. An interesting experiment regarding hydrogen atom production is illustrated in a boron silicate glass radio-frequency electrodeless discharge. Only a small amount of dissociation, on the order of 10%, is obtained from H_2.

In the presence of a small percentage of O_2, however, H_2 dissociation increases to about 50%. But if O_2 is introduced to the discharge first, then removed after a few minutes, and then followed by H_2 alone, the dissociation of H_2 reaches 90% or higher and continues to operate for several hours at this high degree of dissociation. It appears that oxygen in this discharge causes the walls to be activated in some way, so that they lose their ability to recombine the atoms back into molecules. Fite suggests that the lack of understanding of adsorption of gas-discharge products by discharge reactor walls and the catalysis of reactions by these walls is a major stumbling block in understanding the chemistry that occurs in discharges.

An important macroscopic consequence of a plasma state is the shielding that it provides against electrical fields. With the high degree of mobility of electrons and ions in a plasma, field chiefly due to polarization of plasma cannot be applied. For a plasma of $T_e = 2.5$ eV and a number density of 10^{10} cm^3, the Debye length is approximately 10^{-2} cm. The polarization of plasma greatly inhibits the electrical fields in the interior discharge and is the cause of the weak-field condition that exists in the interior of r.f. discharges. Because of these small fields within a well-developed plasma and the fact that ions can easily transfer kinetic energy to neutral molecules, the kinetic temperatures of ions and neutrals should be expected to be very similar and only slightly higher than the temperature of the walls of the discharge. It has been shown that the electron temperatures can be high, even in quite weak fields, because of the poor energy exchanges between electrons and heavy particles.

Afterglow occurs either when the electrical power is removed from a discharge in a closed discharge tube or when gas is permitted to flow in a rapidly evacuated tube away from the region where the electrical power is applied. In both cases the electrons will cool in elastic collisions with the neutral gas, with temperature decay times on the order of 10 μsec. Excited states will also decay, sometimes on the order of msec, leaving only those species that have slow decay constants. Ions and electrons are among the last species to disappear by recombination and ambipolar diffusion to the walls.

Investigations of the afterglow have been a major source of information on discharge chemistry. In the absence of electron-impact ion production processes, the separate charge transfer and ion-molecule reaction processes can be illustrated by mass spectrographic monitoring of the afterglow. In this way, most of the thermal energy ion-neutral reaction data have been obtained [7]. Furthermore, the flowing afterglow has significantly contributed to the knowledge about free radical reactions [8].

4. Bimolecular Discharges

A. General

At the onset of a reaction, discharge results from randomly existing free electrons that are accelerated in the electrical field until they accrue sufficient energy to induce ionization of some of the gas molecules. Electrons generated in this ionization are in turn accelerated to produce additional ionization. As this accumulative effect causes extensive breakdown of the gas the current increases and the discharge is established. A steady state will be reached in which an equilibrium exists between the rate of formation of ions and the rate of their recombination.

The higher the field strength, the more energy the electrons can pick up between collisions, that is, an increase in field strength raises the average electron energy. This results in a greater number of inelastic collisions and thus increases the chemical field. It is to be noted that an increase in power after ignition enhances the degree of ionization and thus reduces the resistance. Therefore, the field strength in glow discharges changes only slightly over a large range of power settings. An increase in pressure causes a reduction of the average mean free path of the electrons. Consequently, they pick up less energy between collisions. Therefore, an increase in pressure will shift the electron-energy distribution to lower values and will thus reduce the rate constant for inelastic collisions.

In bimolecular processes, an additional effect is observed. The yield first rises with increasing pressure due to the higher rate of reaction and then drops because of the lower electron energy [3].

B. Free Radicals

A considerable body of analytical evidence has shown that a wide variety of free radicals are formed in discharges. Those first identified included CH, OH, CN, CS, $R-CH$, OH, CNO, CNS, CF, CF_2, C_6H_5, NH_2, PH, PH_2, SH, S_2H, and others. A free radical is most often formed from gas molecules by the abstraction of an atom. A common example is the breaking of an H atom from hydrocarbons; others include the abstraction of halogen, sulphur, and oxygen from their molecular bodies, producing radicals such as CH_3, C_2H_5, C_6H_5, CCl_3, CS, and OH. Free radicals are also produced by breaking off more than one atom, that is, $NH_3 \rightarrow NH$ or $CH_4 \rightarrow CH$, or by rupturing molecular chains or producing CH and CH_2 radicals from hydrocarbons containing many C atoms.

Free radicals are also generated in a glow discharge from gas mixtures; for instance, NH can be found in nitrogen and hydrogen mixtures or OH in oxygen and hydrogen mixtures. The energies required for such free-radical formation lie well within the eV ranges in normal glow discharge. For simple radicals the collisional efficiency of recombination is typically close to unity—nearly every collision will result in recombination. Three body collisions are of lesser importance. The rate constants for the recombination of radicals such as CH_3, C_2H_5, and CF_3 are all in the vicinity of 2×10^{10} 1 · mole^{-1} sec^{-1}, with activation energies close to zero.

Data from thermal decompositions of hydrocarbon compounds indicated that the hydrocarbon radicals recombine at similar rates and that their half-lives are on the order of about 10^{-5} sec. Steric effects are known to cause the half-lives to increase considerably, however. Like radicals can combine by disproportionation:

$$2\ C_2H_5 \rightarrow C_2H_4 + C_2H_6,$$

or by straight combination:

$$2\ C_2H_5 \rightarrow C_4H_{10}$$

at approximately similar rates.

The considerable number of different types of radicals that exist in a plasma may result in a large variety of recombination products. Abundant evidence exists to show that when oxygen or hydrogen atoms impinge on reactor walls or on particles of silica, carbon, or other materials, such materials will frequently become incandescent. It is important that such localized heating, or any heating, be controlled to inhibit the occurrence of ordinary thermal reactions in such interchanges. As an example, in the reaction between carbon and atomic oxygen, heating of the carbon could cause it to burn in molecular oxygen. The fraction of atoms lost via collision at various surfaces at pressures of approximately 0.1 torr is shown in table 3.1.

Table 3.1. Fraction of O, H, and N atoms lost by collision at surfaces

Surface	$\alpha \cdot 10^5 H$	$\alpha \cdot 10^5 O$	$\alpha \cdot 10^5 N$
Silica	70	60	—
Pyrex glass	46	3[a]	1.5–3
PbO	—	—	4.5×10^3
CaO	—	1.6×10^2	—
CuO	—	4.3×10^3	—
Pt	—	—	2.5×10^3

[a]HF-treated glass.
Source: Author.

C. Hydrogen

Typical bimolecular discharges include those of H_2 and O_2. When hydrogen gas molecules collide with electrons, the collisions with low-energy electrons will be elastic, but their energy is increased with each collisional step until the collisions cause excitation or breakdown of hydrogen. Although the band energy of H-H is 4.47 eV (or 103 Kcal/mole), electrons having an energy of 4.5 eV will generally not break hydrogen molecules into atoms. At an energy of 9 eV the hydrogen molecule will be excited to its repulsive $^3\Sigma\mu$ triplet state, at which dissociation in two H atoms occurs. In addition to excitations of other electronic levels of the molecule as the electron energy rises, further formations of stable H_2^+ ions at the ionization potential of 15.4 eV may take place.

Electrons of energy higher than that needed to produce a given species may produce that species with correspondingly higher kinetic energy. Furthermore, the atoms derived from dissociation may also be excited by electron collisions to higher states or may form metastable states that decay by further collisions and raise or lower their energy to values that induce radiation. It should be noted that in a high-frequency discharge the distribution of electron energy is closely Maxwellian and that the electrons in the high energy tail of the distribution are those that bring about ionization and excitation.

In addition to these phenomena, hydrogen atoms combining by a three-body collision process can form unstable molecules that retain the greater part of their recombination energy and may survive up to 10^8 collisions in the gas phase. This demonstrates that chemical reactions of active hydrogen can be due to one or more of the following species:

Hydrogen atoms (H), excited hydrogen atoms (H*);
H_2^+ ions, excited hydrogen molecules (H_2*);
Protons (H^+) or unstable complexes (H_x*).

Other gases exhibit a similar behavior. For instance, in oxygen the potential for the O^- ion formed by electron capture:

$$O_2 + e \rightarrow O_2^-$$
$$O_2^- \rightarrow O + O^-$$

is 4.53 eV, that is, on the order of magnitude existing under plasma conditions. At energies of 17.28 eV O^+ and O^- ions can be formed. Hydrogen halides can form atoms and negative ions,

$$HX + e \rightarrow H + X^-$$

at low energies ranging from 1.88 eV for HF to only 0.03 eV for HI. Ion-pair formation together with other mechanisms is most commonly found in the dissociation of the halogens

$$X_2 + e \rightarrow X^+ + X^- + e$$

Thus, electron addition is the process occurring at lowest electron energy level by so-called resonance capture to produce a negative ion:

$$AB + e \rightarrow AB^-$$

Dissociation of this negative ion will result in a neutral atom and a negative ion:

$$AB^- \rightarrow A + B^- \text{ or } A^- + B$$

At higher energies, positive and negative ions may form:

$$AB + e \rightarrow A^+ + B^- + e$$

As the energy is further increased a positive molecular ion results:

$$AB + e \rightarrow AB^+ + 2e$$

This may dissociate to yield a positive ion and a radical or atom:

$$AB^+ \rightarrow A^+ + B$$

It should be noted that such simplified gaseous dissociations do not take into account ionization cross section, charge transfer, and the effects of collisions of positive ions and neutral atoms with the walls of the confining reactor, which may induce secondary electron emission, sputtering (that is, the ejection of atoms or aggregates of atoms), or negative ions.

The observable characteristics of a plasma are functions of the species that constitute the plasma. Collisions between the species play a particularly significant part in ionized gases. For a weakly ionized gas, such as that found in glow discharges, the dominant collisions are those between charged and neutral species and those between neutral species. Since molecular gases have much more frequent electron-impact cross sections than atomic gases, they exhibit different characteristics. Electronegative gases form stable negative ions upon electron impact, while electropositive gases yield positive ions and electrons.

D. Oxygen

Since molecular oxygen is an electronegative gas, it forms stable negative ions. The electrical and optical characteristics of the oxygen-positive column are unusual. They occupy an intermediate place between the normal electropositive discharges, such as nitrogen or rare gases, and the strongly electronegative discharges by iodine.

The electron collisional aspects of H_2, N_2, and O_2, in which elastic and inelastic collision cross sections have been derived by numerical solution of the Boltzmann transport equation, were compared with experimental data by Phelps and coworkers [9,10]. For H_2 and D_2 the effective vibrational excitation cross section of 10^{-17} cm^2 becomes appreciable at 1 eV and peaks near 5 eV at 8×10^{-7} cm^2. For dissociation, a threshold was found at 8.85 eV and peak of 4.5×10^{-17} cm^2 at 16 to 17 eV. For average electron energies of 2 eV to 3 eV, the dissociation constants were calculated to be on the order of 10^{-10} cc molecules^{-1} sec^{-1}.

Both the electron-ion recombinations at the wall and the reaction

$$H_2^+ + H_2 \rightarrow H_3^+ + H$$

are very fast and can equal the ionization rate. The dissociation yield in H_2 discharges is believed to be chiefly controlled by surface recombination. The term "surface efficiency" has been introduced to explain atom recombinations and such phenomena as increased yields in the presence of small amounts of H_2O or O_2 [11]. This latter effect of "catalytic" production of H when small amounts of H_2O or O_2 are present has been shown to be more than tenfold when 0.1 to 0.3% H_2O is added to dry H_2 in radio-frequency discharges.

A very large body of experimental work has been reported on oxygen discharges. In 1927 Guntherschulze [12] observed that the oxygen-positive column existed in two forms, namely, one with a high-voltage gradient and grayish green luminescence and the other with a low-voltage gradient and violet in color. Later they were described as H-form (for high, or *hoch* in German) and T-form (for low, or *tief* in German). The H-form existed at higher currents and lower pressures than the T-form (figure 3.3).

When both forms were present in the column, they were found to be axially separated, with the H portion on the anode side and the T portion on the cathode side. It was observed that the T-form progressed from the cathode to the anode and the H-form proceeded from the anode to the cathode. It is the T-form that is considered to differ from other molecular gases.

Thus the oxygen plasma is a three-component plasma consisting of three types of charge carriers: electrons, positive ions, and negative ions. In the glow discharge the predominant positive ion is the O_2^+ with minor quantities

Figure 3.3. H- and T-forms in oxygen-positive column. *Source:* Adapted from J. W. Dettmer, Ph.D. Dissertation, Air Force Inst. of Tech., 1978.

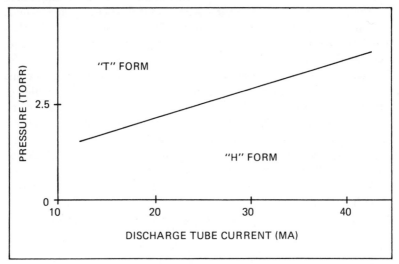

of O^+, while the predominant negative ion was found to be O^-[13] with a 10:1 ratio of O^- to O_2^- ions [14]. In some cases, the negative ion has been found to exceed the electron concentration by up to 20:1 [15]. The dominant ionization and ion processes as well as the major attachment and detachment processes are summarized in table 3.2 [16].

Electronic excitation is described by three processes with threshold at 4.5, 8.0, and 9.7 eV. The latter two are expected to lead to dissociation. At an

Table 3.2. Dominant + ionization, loss, attachment, and detachment processes in oxygen discharges

Ionization	Attachment
$e + O_2^* \rightarrow O_2^+ + 2e$	$e + O_2 \rightarrow O + O^-$
$e + O_2 \rightarrow O_2^+ + 2e$	$e + O_2^* \rightarrow O + O^-$
$e + O \rightarrow O^+ + 2e$	$e + 2O_2 \rightarrow O_2^- + O_2$
$N_- + \text{electrode} \rightarrow N + e$	$e + O_2^* + O_2 \rightarrow O_2^- + O_2$

Nonattachment losses	Detachment
$e + O_2^+ \rightarrow 2O$	$e + O^- \rightarrow e + O_2$
$e + O_2^+ + M \rightarrow O_2 + M$	$O^- + O_2^* \rightarrow O_3 + e$
$e + N_+ + \text{wall} \rightarrow N$	$O_2^- + O \rightarrow O_3 + e$
$e + N_+ + \text{electrode} \rightarrow N$	$O_2^- + O_2^* \rightarrow 2O_2 + e$

Source: J. W. Dettmer, Ph.D. Dissertation, Air Force Inst. of Tech., 1978.

average electron energy of 3.0 eV, the effective dissociation rate constant is approximately 2×10^{-9} cc molecule^{-1} sec^{-1}. The corresponding ionization rate constant is 1.3×10^{-10} cc molecule^{-1} sec^{-1}.

Surface recombination of O atoms is reported to be kinetically similar to that of H atoms except for a lower diffusion coefficient and molecular velocity by a factor of three. In pure O_2, as in N_2, there exists another loss term not observed in H_2. Although not shown in table 3.2, the polyatomic ions O_3^+ and O_4^+ can react exothermically via

$$O_4^+ + O \rightarrow O_3^+ + O \rightarrow O_2^+ + O_2$$

to recombine O atoms. O_4^+ is considered [17] a major ion, that is, $[O_4^+] = 10^{11}$ cm^{-3} and, regenerated from the bimolecular reaction, $O_2 + O_2^+ \rightarrow O_4^+$.

The process $O^- + O \rightarrow O_2 + e$ is known to occur at high rates [18] ($k = 1.5 \times 10^{-10}$) and should follow the dissociation attachment step $e + O_2 \rightarrow L^- + O$ (table 3.1) which has a threshold of 4.5 eV and a low maximum near 7 eV.

E. Nitrogen

In nitrogen discharge the considerable complexity of "active nitrogen" is ascribed to its larger cross sections for vibrational excitation as well as to the existence of metastable electronically excited states below the dissociation limit of ground-state N_2. As a consequence, extensive vibrational excitation continues for much longer times than the time open in the discharge zone, and chemiionization has been detected in the "pink glow" downstream of the discharge. Highly excited ground-state molecules are believed to be the principal carriers of excitation. At an average electron energy of 3 eV in a Maxwellian distribution, the effective dissociation rate constant has been reported by Kaufman [17] to be $k_d = 3 \times 10^{-10}$, with a corresponding ionization rate constant, k_i, of 6×10^{-11}.

The dissociation rate is considerably lower than that of H_2 or O_2, which explains the difficulty of obtaining extensive dissociation of N_2 in glow discharges. Loss processes due to atom recombination should be similar in magnitude to those of oxygen, since the molecular velocities are similar. A catalytic atom-loss mechanism

$$N_4^+ + N \rightarrow N_3^+ + N_2$$
$$N_3^+ + N \rightarrow N_2^+ + N_2$$
$$N_2^+ + N_2 \rightarrow N_4^+$$

has been proposed by Young [19]. The binding energy of N_4^+ has been determined to be about 0.5 eV [20], but it directs bimolecular formation to be slow [21].

Winkler [22] appears to be the first to have obtained a high level of 30% N atoms when using moist N_2 in a pulsed condensed discharge of 4–5 pulses per sec. Fontana [23] observed later that the N atom concentration in the discharge was increased by the presence of traces of water vapor. Young and colleagues [24] found an increase of N atom concentration from pure N_2 from 2×10^{13} to 1.2×10^{15} atoms/cc by the addition of 1% of SF to N_2 prior to discharge. Both NO and O_2 resulted in similar increases when added to N_2 in larger quantities.

For homogeneous recombination reactions:

$$N + N + M \rightarrow N_2 + M$$

Rabinowitch [25] had, as early as 1937, made an estimate for the rate constant to be $K_1 = 1.10^{10}$ 1^2 mole^{-1} sec^{-1} by correctly assuming that the afterglow was due to N atom recombination. This figure is well in agreement with the currently accepted values. The intensity of the afterglow is now known to be proportional to the rate of disappearance of N atoms and not strongly dependent on the nature of a third body [26] when M is N, N_2, Ar, or He. The processes involved in the afterglow, following the initial atom recombination, must involve excited N_2 [27].

F. Summary

Electron-impact excitation of diatomic gases is of particular importance in those molecules that have large cross sections, such as N_2, CO, and CO_2, but is considerably less so in H_2 and O_2, whose cross sections are smaller by a factor of 10 to 50 than those in N_2. The high vibrational temperature, T_v, of N_2 in glow discharges is generally believed to be responsible for much of its electronic excitation. Thus, if T_v is close to the electron temperature, T_e, the internal equilibrium of vibrational states is liable to amount to higher concentrations of molecules with 5 to 8 eV of vibrational energy.

Direct dissociation of a diatomic molecule is caused chiefly by impact of electrons whose energy is larger than the dissociation energy and can occur, depending on the particular mechanism, above 10^{-13} sec.

The principal processes can be summarized as follows:

Dissociation by electron impact via excited states is chiefly dependent on the existence of a dissociating state at adequate excitation energy. If the latter is small, such as in H_2 and O_2, the rate will be high. For N_2, whose dissociation energy is larger, the rate will be about 10 times smaller.

Electron-ion recombination at the wall or in the gas will also generate atomic species in reactions, such as $e + O_2^+ \rightarrow 2O$, $e + N_2^+ \rightarrow N + N$, and so forth. The upper limit to this atom-generation term is given by the total rate of ionization except for a factor from the stoichiometry of the dissociation.

Ion-molecule reactions of primary ions may occasionally be sufficiently exothermic to generate atoms and ions of lower ionization potential, such as in $H_2^+ + H_2 \rightarrow H_3 + H$; however, the same reactions for O_2 and N_2 are endothermic.

Neutral-neutral dissociation may also take place. Since the kinetic temperature of neutral species is low in comparison to that of the electrons, two excited molecules would have to react to transfer their excitation to a predissociating state. Such reactions are believed to occur less frequently.

The aforementioned processes are sometimes referred to as "production terms," while loss terms are distinguished by processes such as:

Homogeneous gas-phase atom recombinations, which are negligibly small and have rate constants in the order of 10^{-32} cm^{-3} molecule^{-1} sec^{-1}.

Surface recombinations, on the other hand, which are well known and are defined by a rate constant, $k = \gamma c/d$, where d is the diameter of a cylindrical tube, γ is the recombination coefficient, and c is the average molecular velocity of the atoms.

Ion-molecule reactions, capable of removing atoms from polyatomic ions, such as $N + N_3^+ \rightarrow N_2 + N_2^+$.

Neutral, bimolecular atom-molecule reactions, such as $N + O_2 \rightarrow NO + O$ or $O + H_2 \rightarrow OH + H$, which frequently exhibit considerable activation energies.

5. Reactions of Diatomic Gases

A. General

It has been noted in this chapter (section 4b) that a large number of free radicals form in glow discharges. Ionization potentials known for many free radicals can be used for the calculation of bond dissociation energies when the so-called appearance potential of the ionized free radicals can be measured. The appearance potential is equivalent to the heat of formation of the reaction:

$$e + RH \rightarrow R^+ + H + 2e$$

and the ionization potential is equivalent to the heat of reaction:

$$R \rightarrow R^+ + e$$

Table 3.3. Bond dissociation energies $R(CH_3 - X)$

Species X	Electron impact (eV)	Other methods (eV)
H	4.43	4.37
CH_3	3.99	3.60
ND_2	2.54	2.34
OH	3.95	3.90
Cl	3.40	3.47
Br	2.33	2.91
I	2.30	2.30

Source: W. L. Fite, in *Chemical Reactions in Electrical Discharges,* ed. B. D. Blaustein, Advances in Chemistry Ser., *No. 80,* Amer. Chem. Soc.: Washington, D.C. (1969).

The heat of the reaction is obtained by combining these:

$$RH \rightarrow R + H$$

Electron-impact studies for a number of simple molecules have resulted in bond energy values that are in good agreement with values obtained from other methods (table 3.3).

The aforementioned dissociations have led to the assumptions that bond energy values derived from thermochemical data can frequently be applied in the formation of free radicals and that free radicals can be produced in glow-discharge plasmas, since the energy required lies well within the range of that existing in such discharges. In this way calculations have provided data for the formation of ions of the type $(CH_3^+ - X)$ and for ions resulting from the removal of more than one atom from the molecule. Since energies of a large number of bonds, that is, the carbon-carbon bond in compounds such as CH_3-CH_3, CH_3-CN, C_6H_5-CH_3, CH_3CO-CH_3, and many others, as well as C-Br, C-S, C-H, and CO-O rarely exceed 5 eV, a sizable number of various free radicals with some ions is expected to occur in glow discharges.

B. Reactions of Hydrogen

The energy of hydrogen atoms is too high to permit recombinations on collision. In order to remove this energy, a third body must be present. Therefore, such recombinations can occur at the walls of the reaction vessel. Traces of water vapor were also found to enhance the rate of homogeneous recombinations:

$$H + O_2 + M \rightarrow HO_2 + M$$
$$H + HO_2 \rightarrow H_2 + O_2$$

or

$$H + HO_2 \rightarrow 2\ OH$$

or

$$H + HO_2 \rightarrow H_2O + O$$

Transfer reactions of the type:

$$H + H_2 \rightarrow H_2 + H$$

have also been measured, since the ortho and para forms of hydrogen can be distinguished. Similarly, transfer reactions of H and D atoms with hydrocarbon,

$$CH_4 + H \rightarrow CH_4 + H$$
$$CH_4 + D \rightarrow CH_3D + H$$

have been found to take place above 200°C. Unsaturated hydrocarbons are also known to react with atomic hydrogen, generally by addition in the first step; thus propylene may yield via two steps

$$H + C_3H_5 \rightarrow C_3H_6$$

and

$$H + C_3H_6 \rightarrow C_3H_7$$

while ethylene may form two methyl radicals:

$$H + C_2H_4 \rightarrow C_2H_5$$
$$H + C_2H_5 \rightarrow 2\ CH_3$$

Acetylene has a tendency to transfer deuterium:

$$D + C_2H_2 \rightarrow HD + C_2H$$
$$D + C_2H \rightarrow C_2DH$$

Hydrogen has also been found to react with compounds of carbon, hydrogen, and oxygen in a variety of ways:

$$H + HCOOH \rightarrow H_2 + COOH$$

and

$$H + CH_3OH \rightarrow H_2 + CH_2OH$$

and

$$H + CH_3CHO \rightarrow H_2 + CH_3CO$$
$$H + CH_3CO \rightarrow CH_3 + CHO$$
$$H + CH_3CO \rightarrow CH_4 + CO, \text{ and so forth}$$

and

$$H + CH_3OCH_3 \rightarrow H_2 + CH_3OCH_2$$
$$H + CH_3OCH_2 \rightarrow CH_3 + CH_3O, \text{ and so forth}$$

The reaction products from ($H + CH_3COCH_3$) were found to be CO and CH_4.

Hydrogen atoms are known to be not only highly reactive but also strong reducing agents. Thus, many oxides, including those of copper, bismuth, lead, silver, mercury, and tin, are generally reduced to metals.

Likewise, halocarbons and alkyl halides are found to be reduced by hydrogen atoms:

$$H + CCl_4 \rightarrow HCl + CCl_3$$
$$2 CCl_3 \rightarrow C_2Cl_6$$

Carbon monoxide and H form primarily formaldehyde.

Reactions of H with inorganic halides may result in the abstraction of a halogen, such as in the formation of $TiCl_3$, $ZrCl_3$, $ZrBr_3$, $ZrIr_3$, and VCl_3 from their tetrahalide. Other halides, such as BCl_3, $AlCl_3$, AlI_3, AlF_3, $ScCl_3$, ScF_3, $AsCl_3$, $SnCl_4$, and $SICl_5$ were generally found to be reduced to the metals when the hydrogen was mixed with the halide vapor prior to passing through the discharge. Hydrogen will also form hydrides under glow-discharge conditions. Thus, hydrides of phosphorus, sulphur, arsenic, antimony, sodium, and silver have been isolated under appropriate conditions. Likewise, H atoms react with halogens to produce the compounds HX followed by chain reactions of the type

$$H + Cl_2 \rightarrow HCl + Cl$$

and

$$H + HX \rightarrow H_2 + X$$
$$X + H_2 \quad HX + H$$

Methane and ethane were obtained by reaction between carbon and hydrogen downstream away from the luminous region of the discharge.

Oxygen reacts with hydrogen at ambient temperatures to form primarily water. H_2O_2 was found when the reaction took place below $-80°C$, however. On the other hand, N_2 molecules do not react directly with H atoms but will form NH_3 from N_2H_4 at elevated temperatures. (SO_2 + H) is found to yield H_2S and H_2O; NO_2 will be reduced by H to NO to react further to form HNO, which will further decay in the presence of H to NO and N_2O.

The formation of CH_4 and C_2H_2 was reported [28] to occur from H_2 + CO mixtures in microwave discharges at about 20% conversion. Conversion of CO was increased to over 90% when the reaction products were removed by a cold trap as they formed. With H_2 + CO_2, no hydrocarbons were formed unless water was frozen out.

Reduction of metal oxides in hydrogen plasma has recently been reported [29]. Thus, CuO, NiO, PbO, and ZnO were reduced to their metals in a plasma generated at 8 MHz in the pressure range 10–50 torr and power levels of 2–3 kW r.f. Aluminum oxide was not appreciably reduced in this type of discharge. Concentrations of atomic species in diatomic molecular gases are well in excess of the equilibrium values predicted at the kinetic gas temperatures of 700–2000°K prevailing in these discharges, and a strong disequilibrium exists between electronic, vibrational, rotational, and translational temperatures.

C. Reactions of Oxygen

In a homogeneous recombination reaction of atomic oxygen in a glow discharge at ambient temperatures, ozone will be formed in small concentrations that will decay to molecular oxygen at relatively high rates. In these reactions:

$$O + O_2 \xrightarrow{K_1} O_3 + M \ (M=O_2)$$

$$O + O_3 \xrightarrow{K_2} 2O_2$$

and

$$O + O + M \xrightarrow{K_3} O_2 + M$$

The values for K_1, K_2, and K_3 were found to lie in the following ranges:

$$K_1 = 0.6 \text{ to } 2.0 \times 10^8 l^2 \text{ mole}^{-2}\text{sec}^{-1}$$
$$K_2 = 3 \times 10^{10} \text{ to } 4.3 \times 10^9 \text{ l mole}^{-1}\text{sec}^{-1}$$
$$K_3 = 1 \times 10^9 l^2 \text{ mole}^{-2}\text{sec}^{-1}.$$

At very low temperatures a conversion of up to 78% ozone has been obtained when condensing on a surface was followed by warming. It is believed that the ozone is formed during condensation by a reaction of O and O_3 at the surface. Because of the industrial importance of ozone areas such as water purification, air cleansing in ventilation systems, and bleaching, a variety of ozonizers have been developed for commercial applications. These exhibit high efficiencies even when operating with air at atmospheric pressures, although higher yields are obtained from those using O_2.

Hydrocarbons may be attacked in glow discharges by O atoms via different reactions: by hydrogen abstraction, by addition to a double bond, and by entry into a C-H or C-C bond. Hydrogen abstraction appears to be dependent on temperature. Thus CH_4 will be oxidized to the extent of only 1% at room temperature but to 6% at 190°C or higher. Ethane and butane, benzene, and toluene react more than a hundred times slower than the olefins. The initial reaction step is believed to be

$$O + CH_4 \rightarrow OH + CH_3.$$

followed by the well-known fast reactions between O and radicals.

Olefins undergo rapid reactions believed to occur in the following steps:

$$O + C_2H_2 \rightarrow [C_2H_2O] \rightarrow CO + CH_2$$

It is generally assumed that the main reaction between olefins and O is the addition of O to the double bond, thus forming epoxides and carbonyl compounds. The addition operates preferentially at the less substituted carbon atoms of the double bond, forming probably a biradical. In smaller molecules, such as ethylene and propylene, subsequent fragmentation may take place more readily than in larger molecules, which have more degrees of freedom to dissipate the energy generated.

In reactions of O with organic compounds containing oxygen, it is generally found that they yield CO, CO_2, H_2O, and H_2. Hydrogen abstraction appears to be the first step of attack on compounds composed of C, H, and O. Sulphur reacts quickly with O to form SO_2 and SO_3. In the presence of sulphur vapor, SO_2 may also react to form S_2O, which may be stored below 40 mm Hg for several hours. Above this pressure it will decompose to S and SO_2. Chlorine is known to react with O to form initially $ClO + Cl$ if present in a tenfold excess. If O atoms are dominant, Cl atoms may also be produced together with ClO_2 and ClO_3 at megacycle frequencies. Bromine has been oxidized in plasma at 83°K to BrO_2 at a very high yield and in a mixture of O_2, N_2, and Br_2 the compound $BrO_2 \cdot 3NO_3$ was found to condense on the walls of the reactor. Under similar conditions iodine was oxidized to I_2O_2 as the sole product. In reactions between fluorine and oxygen, compounds such as O_2F_2,

O_3F_2, and O_4F_2 have been synthesized. In exposing most metals to atomic oxygen, the corresponding oxides are formed readily. At low O concentrations this reaction may occur at lower temperatures, such as at 200°C for ZrO_2. Carbon and carbonaceous substances react readily with oxygen at activation energies of about 10 Kcal/mole in the temperature range 14–200°C.

The practical aspects of such reactions lie in the utilization of plasma for low-temperature "ashing," that is, converting carbon and carbonaceous material into gases such as CO, CO_2, H_2O, NO, and NO_2. The ash remaining after burning is completed consists of those elements or compounds are volatilized. This technique is useful in a number of applications:

Biological: Organic matter is oxidized without volatilizing trace elements and decomposing inorganic structures for analysis in clinical chemistry, medical research, nutrition, biochemistry, toxicology, and forensic medicine.

Geological: Organic matter is removed without altering crystalline structure.

Filter analyses: Particles and volatile trace elements are separated from filter matrices for study of air pollution and fallout.

Surface activation: Removal of organic coatings from substrates, cleaning surfaces of electronic components, treatment of polymer and metal surfaces to improve adhesion and wettability.

Oxidation of polymer surfaces will be discussed in greater detail in Chapter 8. Surface cleaning will also be dealt with under this heading in the appropriate chapter.

Another potentially important application of plasma oxidation is the removal of trace contaminants from spacecrafts, submarines, and other closed environments by oxidizing the contaminants to produce CO_2, N_2, and water. The contaminants may range from metabolic products, such as alcohols, esters, hydrogen sulfide, and ammonia, to lubricant solvents, such as Freons, aromatics, alcohols, and ketones. In tests on three prototype models, the concentration of methane was reduced by three orders of magnitude, with the discharge operating at pressures from 50 to 300 torr. The removal rate was found to be independent of the concentration of the contaminant, an important advantage over other air-purification systems. The system was developed at the Ames Research Center of the National Aeronautics and Space Administration.

D. Reactions of Nitrogen

The basis for the titration of N atoms is the high-rate reaction:

$$N + NO \rightarrow N_2 + O \qquad (11)$$

This reaction is of technical importance, since the concentration of N atoms determines the yield of a number of compounds made from it, such as NH_3, nitric acid, hydrogen cyanide, and others. Reactions leading to such compounds are generally endothermic:

$$\tfrac{1}{2} N_2 + \tfrac{1}{2} O_2 \rightarrow NO \qquad \Delta H_{298} + 21.6 \text{ Kcal/mole} \qquad (12)$$

$$CH_4 + \tfrac{1}{2} N_2 \rightarrow HCN + 3/2 H_2 \qquad \Delta H_{298} = 49.1 \text{ Kcal} \qquad (13)$$

$$2C + N_2 \rightarrow C_2N_2 \qquad \Delta H_{298} = 71.1 \text{ Kcal} \qquad (14)$$

Therefore, they are favored by high temperatures. In the presence of ozone and an excess of nitrogen, the NO formed reacted according to reaction 11 to give N_2 and O; when ozone is in excess,

$$O_3 + NO \rightarrow NO_2 + O_2 \qquad (15)$$

reaction 15 is dominant, with N_2O present in smaller amounts. For the formation of NO by reaction 12, a high yield of 15.5% was obtained and was shown to be a function of current and pressure. Under 2450 Mc microwave conditions, a high yield of 0.77 moles/kWh was reported [30].

The production of hydrogen cyanide is of importance, since over 50% of it goes into the manufacture of acrylonitrile. The conventional processes involve a reaction between NH_3 and CH_4 over a platinum catalyst at 1300° to 1600°K or at about 1800°K in a fluidized carbon bed at a yield of 85% to 90% of theoretical. The industry has recognized its interest in plasma synthesis of HCN from nitrogen as raw material. High yields were recently reported [31] from the reaction of N_2 and CH_4 in a continuous gas discharge at 35 MHz and r.f. power levels of 2–5 kW in the pressure range of 5–30 torr.

Considerable interest was evoked by the reported reaction of N at polyolefin surfaces [32], which yielded 34% HCN at 89°C. In general, reactions of atomic nitrogen with organic compounds result in their destruction, leaving mainly HCN, some cyanide and NH_3, and low-molecular-weight gases, in addition to polymers [33]. Few cases have been reported where nitrogen is incorporated into the products in reactions such as with butadiene:

$$H_2C\!\!=\!\!CH\!\!-\!\!CH\!\!=\!\!CH_2 + N \longrightarrow$$

$$\underset{H}{\underset{|}{N}}$$

$$CH_3\!\!-\!\!CH\!\!=\!\!CH\!\!-\!\!CN$$
$$C_4H_7CN$$
$$C_5H_9CN$$

When nitrogen and hydrogen mixtures are passed through a discharge, NH_3 with smaller quantities of N_2H_4 is formed from the resulting atomic species. Highest yields of NH_3 are obtained at low temperatures at which the reaction rate is slow.

Reactions of atomic nitrogen with sulphur compounds such as CS_2, H_2S, S_8, OCS, S_2Cl_2, and SCl_2 yielded sulphur-nitrogen structures. These reactions were accompanied by a blue flame due to emission of excited S_2 molecules. The reaction products with sulphur were mainly S_4N_4, S_4N_2, and $(NS)_x$. CS_2 yielded chiefly $(NS)_x$, while H_2S gave compounds such as $(NS)_x$, H_2, NH_3, S_7NH, and sulphur. Metal nitrides are formed with a number of metals. These industrially important reactions will be discussed in greater detail in Chapter 8.

In a recent investigation it was demonstrated [34] that nitrogen can be incorporated into a plasma-polymerized film by using an ethylene-nitrogen gas mixture. Control was provided by the added degree of freedom of the nitrogen partial pressure in the starting gas mixture. This parameter proved to have a large effect on many of the film properties. It has been suggested that this procedure is applicable to a number of gas mixtures. Hollahan and Stafford demonstrated [35] that nitrogen-hydrogen mixtures (as well as NH_3) can add NH_2 groups to various polymer surfaces. These reactions will be discussed in the chapter dealing with surface modifications. At about the same time, Hollahan and McKeever [36] reported on the synthesis of polymers from the inorganic gas reactants CO, H_2, and N_2 in r.f. glow discharges at 13.56 MHz, under 0.3–2.5 torr pressures. Elemental analysis disclosed that those polymers contained C, H, O, and N in various proportions, depending on the experimental conditions, with nitrogen constant, ranging from 0.2% to 17.64%; the proportions increased as the nitrogen flow rate was enhanced. These results offer interesting speculations for the synthesis of compounds such as amino acids and related structures. Indeed, molecules of biological significance were obtained by the action of electric discharges aimed at simulating the presumed primitive atmosphere of the earth [37]. Experiments aimed at synthesizing hydrazine from ammonia in silent discharges have been going on since 1911 [38]. It was not until the early sixties and the quick removal of the hydrazine by a liquid absorbent [39] that this process was considered to be economical. It was observed that the yield varies inversely as an exponential function of the power density at 100 mm Hg. The effect of pressure also follows a negative exponential variation, both effects leading to the general relationship:

$$Y = a \exp(-bP - c\pi) \qquad \text{(Eq. 3.6)}$$

where Y = grams of hydrazine per kWh and π = kW per cc of reactor volume. The most notable feature in this reaction is the progressive increase in hydrazine yield by the use of a liquid absorbent and by pulsing the discharge. With yields of 20 g/kWh, this process became economically attractive.

E. Reactions of Halogens

Experiments in the thirties disclosed that halogen atoms in discharge tubes recombined quickly on the walls of the reactors. Thus, chlorine atoms survived fewer than 20 collisions, bromine atoms recombined after almost every collision, and iodine atoms could not even be detected because of their high rate of recombination in the gas phase. In the early sixties, however, it was observed [40] that certain chemicals such as H_3PO_4, H_2SO_4, H_3AsO_4, H_3BO_3, and $HClO_4$ when applied to the walls caused marked reductions in the recombination rates of these atoms. In more recent times a number of recombination rates have been made, resulting in approximate values of 2×10^{10} $l^2 mole^{-2} sec^{-1}$, bromine 4.9×10^{10} $l^2 mole^{-2} sec^{-1}$, and iodine 1.8×10^{12} $l^2 mole^{-2} sec^{-1}$. The reactions of halogens with hydrocarbons and other organic compounds have early been regarded [41] as particularly important, since they supply a series of activation energies ranging from 0 to 33 Kcal/mole. Three main reactions are distinguished:

$$\text{H abstraction: } X + RH \rightarrow HX + R \qquad (16)$$

$$\text{X abstraction: } X + RX \rightarrow X_2 + R \qquad (17)$$

Double bond addition:

$$X + C \rightarrow CX \qquad (18)$$

Methane was early found to react rapidly in an electrodeless discharge at low pressures with Cl atoms and Cl_2, according to:

$$Cl + CH_4 \rightarrow CH_3 + HCl \qquad (19)$$

$$CH_3 + Cl_2 \rightarrow CH_3Cl + Cl \qquad (20)$$

$$CH_3Cl + Cl \rightarrow CH_2Cl + HCl, \text{ and so forth} \qquad (21)$$

finally yielding CCl_4 and various chlorinated methanes. Bromine proceeds in the same manner.

The reaction with ethane involves a similar chain mechanism. The activation energy varies considerably, however; with Cl it is exothermic, with Br slightly exothermic, and with I endothermic. Such reactions with F are extremely rapid and exothermic and therefore very difficult to study.

Fluorine is generally assumed to abstract H atoms with zero activation energy and to react with H donors at nearly every collision. Hydrocarbons are normally not observed to lose H atoms by reaction with I atoms, however.

Halogen abstractions are observed to occur:

$$Cl_2 + CCl_3Br \rightarrow BrCl + CCl_4 \tag{22}$$

or

$$Br + CCl_3Br \rightarrow CCl_3 + Br_2 \tag{23}$$

followed by

$$Br_2 + CCl_3 \rightarrow CCl_3Br + Br \tag{24}$$

Reaction 24 may occur with acetylene with Br or Cl:

$$Br + C_2H_2 \rightarrow C_2H_2Br_2 + Br \tag{25}$$

Ethylene reacts at low temperatures by addition:

$$Cl_2 + C_2H_4 \rightarrow C_2H_4Cl \xrightarrow{\;Cl_2\;} C_2H_4Cl_2 + Cl \tag{26}$$

Benzene hexachloride was obtained by passing benzene vapor and chlorine through a silent discharge.

F. Reactions of Free Radicals

Free radicals may be generated by electron impact in an electric discharge, thermally, or by photolysis. In a glow discharge, energy is transferred to the neutral molecules by inelastic collisions with electrons, with the result that such molecules may be excited or ionized. Recombination of positive ions and electrons gives rise to highly excited states that may dissociate into neutral fragments but more probably dissociate into radicals.

Atomic hydrogen also abstracts other atoms (commonly hydrogen) from saturated compounds to produce free radicals [42,43] (reaction 27), while with unsaturated compounds it is added and thus also forms a free radical (reaction 28) [44,45].

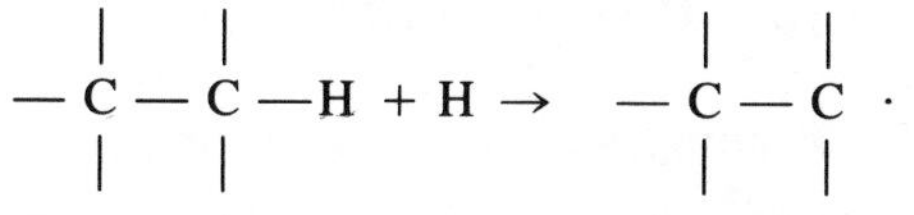

$$-\overset{|}{\underset{|}{C}} - \overset{|}{\underset{|}{C}} - H + H \rightarrow -\overset{|}{\underset{|}{C}} - \overset{|}{\underset{|}{C}} \cdot \tag{27}$$

$$\overset{\backslash}{\underset{/}{C}} = \overset{/}{\underset{\backslash}{C}} + H \rightarrow H - \overset{|}{\underset{|}{C}} - \overset{|}{\underset{|}{C}} \cdot \tag{28}$$

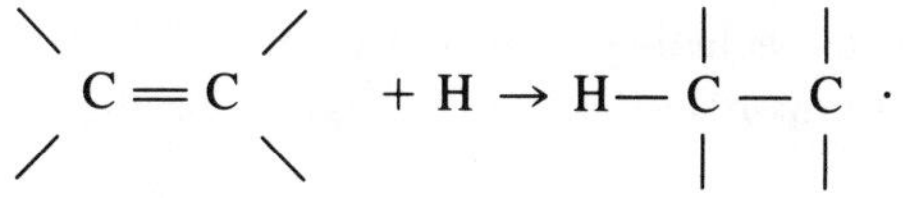

Atomic oxygen reacts with saturated compounds also by abstraction of atoms and the formation of radicals [46].

Hydrogen can be abstracted by a radical R directly:

$$R + HX \rightarrow RH + X \tag{29}$$

where X is H, F, Cl, Br, OH, CH_3, C_2H_5, and so forth. The reaction of CH_3 with toluene can result in hydrogen abstraction both from the methyl group and also directly from the aromatic ring whereby two CH_3 radicals were consumed for each CH_4 formed. Therefore, the mechanism in these two reactions is believed to be:

$$CH_3 + C_6H_5CH_3 \rightarrow CH_4 + C_6H_5CH_2 \tag{30}$$

$$CH_3 + C_6H_5CH_2 \rightarrow C_6H_5CH_3 \tag{31}$$

and

$$CH_3 + C_6H_5CH_3 \rightarrow CH_4 + C_6H_4CH_3 \tag{32}$$

$$C_6H_4CH_3 + C_6H_5CH_3 \rightarrow 2\ C_6H_5CH_2 \tag{33}$$

$$CH_3 + C_6H_5CH_2 \rightarrow C_6H_5C_2H_5 \tag{34}$$

Interestingly, a CH_3 radical may also react with methanol in two ways, namely:

$$CH_3 + CH_3OH \rightarrow CH_4 + CH_3O \tag{35}$$

and

$$CH_3 + CH_3OH \rightarrow CH_4 + CH_2OH \tag{36}$$

whereby reaction 36 is found to be slightly greater than reaction 35. Still, with deuterium in the reactions

$$CH_3 + CD_3OH \rightarrow CH_4 + CD_3O \tag{37}$$

$$CH_3 + CD_3OH \rightarrow CH_3D + CD_2OH \tag{38}$$

the reaction for 37 was nearly four times that for 38 [47].

Hydrocarbon radicals are also found to abstract halogen atoms from a large variety of compounds including CCl_4, C_2Cl_6, and $(CCl_3)CO$.

One of the most important reactions in organic synthesis and polymerization is the well-known addition of an alkyl or other radical to unsaturated hydrocarbons such as

$$CH_3 + C_2H_4 \rightarrow n\text{-}C_3H_7 \tag{39}$$

and the propyl radical then reacts with another ethylene to form the n-pentyl radical and so forth, or it reacts in a chain ending with another radical as is typified in common free-radical polymerization processes.

Other reactions of alkyl radicals include the rapid reaction with NO:

$$CH_3 + NO + M \rightarrow CH_3NO + M \tag{40}$$

(where M may be an acetone). The nitromethane formed may continue to react either with CH_3 or NO.

Methylene CH_2 radicals react with carbon-carbon double bonds to yield initially a cyclopropane ring that will isomerize under the highly exothermic conditions unless it loses its excess energy by other molecular collision processes. By direct insertion of CH_2 into the H-H bond at room temperature, methane is formed together with some CH_3 radicals. Above 100°C, however, the only product of the reaction appears to be ethane.

CCl_3 radicals combine to form hexachloroethane, as their fluoro homologs form C_2F_6. In the discharges of water vapor OH radicals are known to exist, although the main products of such discharges are H_2 and O_2 at room temperature. The OH radicals decay mainly in the following manner:

$$2OH \rightarrow H_2O + O$$

$$OH + O \rightarrow O_2 + H$$

OH radicals will also react with CO to form $CO_2 + H$.

Considerable interest has been shown in the generation of hydrazine from NH_3. From the data for the three possible dissociations of NH_3, namely:

$$NH_3 \rightarrow NH_2 + H \qquad 110 \text{ Kcal/mole} \tag{41}$$

$$NH_3 \rightarrow NH + 2H \qquad 153 \text{ Kcal/mole} \tag{42}$$

$$NH_3 \rightarrow NH + H_2 \qquad 50 \text{ Kcal/mole} \tag{43}$$

reaction 41, although energetically not as favorable as 43, is thought to be the most likely to occur under discharge conditions. Recent chemical advances on the hydrazine synthesis have been discussed in section D of this chapter.

6. Synthesis of Organic Compounds in Plasma

A. General

In general, plasma reactions are strongly dependent on pressure, electrical field strength, and rate of gas throughput. Experiments are usually carried out at pressures of 0.5 to 5 mm Hg, at power levels of 25 to 250 W, and at flow rates of several cm^3 per second. Plasma is often considered an energy source for the activation of endothermic reactions that can also be carried out by conventional high-temperature techniques. Reactions in a glow discharge are also frequently found to proceed by additional mechanisms or mechanisms different from those observed at high temperatures in the absence of plasma, however. A few reviews of plasma chemistry have been published during the last few years [48,49].

The several hundred plasma reactions studied in recent years make it possible to describe the plasma behavior of several classes of organic molecular structures. Some of these, such as isomerizations, eliminations, dimerizations, and polymerizations, have been investigated in greater detail.

B. Isomerizations

A large number of organic compounds are found to isomerize when passed through glow discharges. In many cases such molecular rearrangements are side reactions of small significance. In other cases, however, isomerization is the predominant reaction (table 3.4). Several isomerizations involve migra-

Table 3.4. Isomerization reactions in glow discharge

Starting material	Conversion rate (%)	Products (%)
trans-stilbene	20	*cis*-stilbene 95
Anisol	67	cresol, *o*:48; *p*:29
Phenetol	~30	ethylphenol, *o*:41; *p*:29
N-propyl-phenyl-ether	~30	propylphenol, *o*:38; *p*:19
		ethylphenol, *o*:1; *p*:0.5
		cresol, *o*:7; *p*:4
1-Naphthyl-methyl-ether	13	methyl-1-naphthol, 2:48; 4:35
2-Naphthyl-methyl-ether	88	1-methyl-2-naphthol 45
Diphenyl-ether	40	hydroxybiphenyl, 2:36; 4:18
		dibenzofurane 9
N,N-dimethylaniline	15	*N*-methyl-toluidine, *o*:28; *p*:15
N-methylaniline	6	toluidine, *o*:28; *p*:6
Cyclooctatetraene	80	styrene 40
Pyrrole	6	*cis-trans*-croton nitrile 57
Quinoline	8	cinnamic acid nitrile 33
		cis:5, *trans*:33

Source: H. Suhr and U. Schücker, *Synthesis,* 431 (1970); H. Suhr and R. I. Weiss, *Z. Naturforsch. 256,* 41 (1971).

tion of substituents such as the rearrangement of arylalkyl-ethers to alkyl-phenols [50]. Anisol is found to isomerize largely to form o- and p-cresol. In cresol methylether the o-alkyl group migrates to the o- and p-position as shown below.

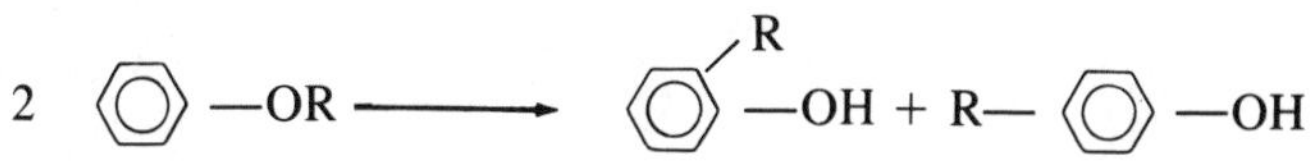

$$2\ \bigcirc\!-\!OR \longrightarrow \bigcirc^{\nearrow R}\!-\!OH + R\!-\!\bigcirc\!-\!OH$$

Similarly, naphthyl methylethers convert readily to methylnaphthols. Related rearrangements are found with the respective nitrogen compounds, which may partly isomerize and partly fragment, as shown below.

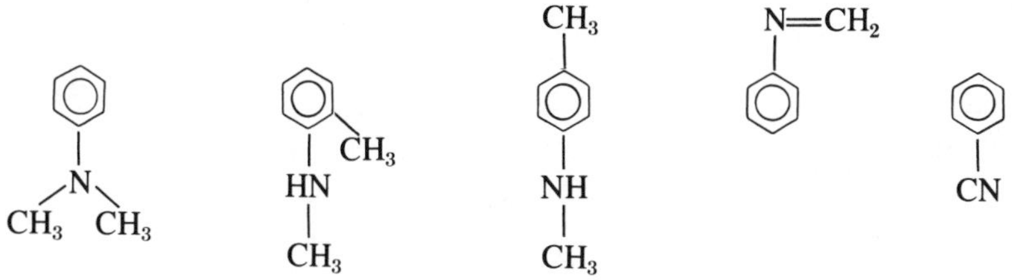

Special cases of isomerization are the pyrrole conversion to croton nitrile, the indol isomerization to benzyl cyanide, pyridine to cyanobutadiene, and quinoline to cinnamic acid nitrile.

C. Eliminations

In many glow-discharge reactions atoms or small molecular entities are separated from their molecules without otherwise affecting the rest of the molecule. For instance, aldehydes are readily decarboxylated to form the respective hydrocarbons [51] (see below).

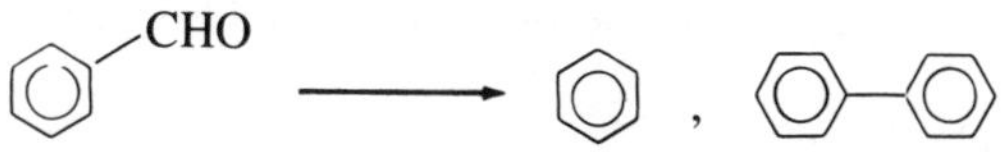

80%   20%

Similarly, pyridine aldehyde yields chiefly pyridin with some dipyridyl. Ketones are also found to decarboxylate without destroying the aromatic base molecule; for example, benzophenone will decompose into diphenyl at high yields. Interestingly, benzoquinone decarboxylates to form predominantly cyclopentadienone [50]. Carboxylic acids are generally found to lose CO_2 without further affecting the parent molecule. Very common in plasmas is the elimination of hydrogen.

Table 3.5. Formation of five-membered rings in glow discharges

Starting material	Conversion rate (%)	Products
⬡—CH₂—⬡	~30	
⬡—CO—⬡	68	
⬡—O—⬡	39	
⬡—NH—⬡	40	
⬡⬡ HO	22	
⬡⬡ H₂N	24	
⬡⬡ O₂N	30	
⬡—NH—⬡ =N	13	

Source: H. Suhr, Chem. Inst. of Univ. of Tübingen, Germany (1973).

Recently, Suhr [52] cited interesting possibilities for syntheses in reactions where elimination results in cyclic products (table 3.5). Six-membered heterocyclic structures have also been plasma-synthesized (table 3.6).

D. Bimolecular Reactions

In addition to the aforementioned monomolecular reactions, bimolecular processes have also been observed. A simple form of bimolecular reaction, for instance, is the dimerization of the starting compound by a loss of hydrogen or another group. Thus, in a plasma, benzene may form biphenyl and sometimes terphenyl [53] (see below). Dimerization of toluene and substi-

Table 3.6. Formation of six-membered rings in glow discharges

Starting material	Conversion rate (%)	Products
CH$_2$—CH$_2$ (two phenyl groups)	~30	(tricyclic product)
CH=CH (two phenyl groups)	77	(phenanthrene-type product)
CH=N (two phenyl groups)	29	CH=N (product)
CH$_3$ | C=N (two phenyl groups)	18	H$_3$C—C=N (product)
N=N (two phenyl groups)	80	N=N (product)
CH=CH (two phenyl groups)	22	(N-heterocyclic product)
NH (with NO$_2$)	90	NH ... O (product)
O (with NO$_2$)	~25	O ... O (product)
NH (with NH$_2$)	93	N ... N (product)
CH$_2$ (with NO$_2$)	~30	CH$_2$ ... O (product)

Source: H. Suhr and U. Schücker, *Synthesis,* 431 (1970).

tuted toluenes is found to yield diarylethanes [54,55]. In general, aliphatic compounds are found to dimerize less readily. Methane will form a variety of saturated and unsaturated hydrocarbons. Propene and substituted propenes, however, will chiefly yield hexadienes via propenyl radicals, $CH_3—CH{=}CH_2$. Isobutylene has been found to form equal proportions of dimethylhexadiene and dimethylhexene.

Most common in organic plasma synthesis are the polymerization reactions. Almost every organic compound can be converted into a polymer under the appropriate conditions of glow discharge. The reaction mechanisms vary greatly with the monomer. Thus, for instance, the fragmentation of phthalic anhydride into dehydrobenzene can result in the formation of polyphenyl [56]. Furthermore, isomerization of some nitrogen heterocyclics yields unsaturated nitriles that polymerize readily [52].

H. Suhr [52] emphasizes that plasma reactions with two or more compounds require exacting feed of all substances to ensure both the correct total pressure and the required component ratio. Because of the complexity of such reactions, few published data are available at this time. In some reactions one component is used to act chiefly as a dilution agent. For example, an inert gas acting as carrier for the monomers will cause a reduction in the number of molecular collisions and in a number of conditions can inhibit polymerization. In other cases, such transport gas molecules in states of low excitation can transfer energy to the monomer by collision. Interactions in plasma between two different molecules have been studied in greater detail. Thus, mixtures of oxygen and benzene yield some phenol, and benzene and ammonia form aniline.

Other mechanisms of bimolecular reactions involve reactive intermediates that are trapped with various additives. The compounds have been found [56] to decompose in two steps to form dehydrobenzene, which subsequently dimerizes in two steps via a biphenyl radical, as shown below.

The intermediates formed in these reactions can react with other molecules present. Hydrogen, for instance, will trap the diphenyl diradical, while acetylene will combine with the dihydrobenzene. Ammonia will form additional compounds with all intermediates to yield aminobenzaldehyde, benzamide, aniline, and carbazol [52].

E. Reaction Mechanisms

Because of the complexity of plasma systems, relatively little is known about the mechanism involved in plasma reactions. The variety of species, such as electrons, positive and negative ions, atoms and molecules at various states of excitement, free radicals and neutral molecules, offer a range of different interactions under glow-discharge conditions.

In most reactions, molecules are first excited through direct collisions, via negative ions, or by the recombination of positive ions with electrons. The excited molecules can fragment, or they can isomerize to form either stable compounds or reactive intermediates. Often, the reactive species, such as the excited molecules, ions, or intermediates, interact with other molecules to form bimolecular structures.

Given the variety of species and the shift in reactivities with change in variables such as concentrations, flow conditions, and species location in the plasma zones, in addition to equipment conditions (wattage, temperatures, preset addition of component molecules, and so forth), it has been most difficult to estimate the importance of the various elementary processes in an environment. Therefore, researchers have usually resorted to speculation about the reaction mechanisms involved from a comparison of plasma products from different reactions. Frequently, similarities are observed between plasma chemistry and photochemistry, pyrolysis, radiochemistry, or mass spectroscopy. In the event that the reaction products are identical, one may assume a similarity of the reaction steps involved. In some cases, the products point to mechanisms that have no known analogues in other areas of organic chemistry [57]. Few mechanisms of plasma reactions have been investigated in greater detail. In such experiments only the starting components and final products are known. Intermediates are rarely identified as to time of existence and concentration. Therefore, many questions regarding the mechanism remain unanswered at this time.

Recent data indicate a large number of applications and insights into mechanistic aspects, however. Most experiments have dealt with simple systems. Suitable models are aromatic compounds because of their considerable stability under plasma conditions. Such experiments are usually limited to the gas phase; that is, to compounds that are vaporous in vacuum without decomposing. If they carry several reactive groups, then they would often yield a number of different products. If one of the possible reaction routes requires substantially less energy than others, however, the reaction would lead chiefly to a single product. It is fortunate that under constant plasma conditions, the result of the reaction can be reproduced with notable accuracy.

7. Synthesis of Inorganic Compounds

Hydrides of boron, silica, germanium, phosphorus, and arsenic can be degraded in glow discharges to reform mixtures of the corresponding higher-molecular-weight compounds. This reaction is an exothermic process. Since reactions of molecular hydrogen are fairly slow at ordinary temperatures, there is no interference with these syntheses. Thus, by passing a mixture of SiH_4 and PH_3 through an ozonizer discharge, a mixture of the compounds SiH_3PH_2, SiH_5PH_2, and $(SiH_3)_2PH$ is formed, together with P_2H_4 and various polysilanes [58]. Higher-molecular-weight hydrides prepared from simple hydrides are listed are listed in table 3.7. The conversion of acetylene-boron hydride mixtures into carboranes involves more complex molecular rearrangements, and a more powerful glow discharge between copper electrodes has been recommended for this process [59]. Volatile halides, such as $SiCl_4$, $GeCl_4$, and BCl_3 have also been converted to higher homologs in a glow discharge. In the presence of suitable reducing agents, such as mercury and copper wool, good yields have been obtained. The products isolated in such reactions include B_2Cl_4, Si_nCl_{2n+2}, Ge_2Cl_6, and P_2Cl_4 [60]. Side reactions that may occur in these syntheses can be avoided to a degree by permitting a stream of atoms or radicals formed in the discharge to impinge on various compounds outside the discharge. Fluorocarbon products such as CF_4, C_2F_4, C_2F_6, C_3F_6, and C_3F_8, as well as fluoronitrogen compounds such as NF_2, NF_3, N_2F_4, and N_2F_2, have been synthesized under plasma conditions that are outside the scope of this book. They include high-intensity arcs, high-voltage discharge, and microwave discharge.

Table 3.7. Plasma synthesis of volatile hydrides

Starting material	Electric discharge	Products isolated
SiH_4	Ozonizer	Si_2H_6, Si_3H_8, Si_4H_{10}, Si_5H_{12}, Si_6H_{14}, Si_7H_{16}, Si_8H_{18}
GeH_4	Ozonizer	Ge_2H_6, Ge_3H_8, Ge_4H_{10}, Ge_5H_{12}, Ge_6H_{14}, Ge_7H_{16}, Ge_8H_{18}, Ge_9H_{20}
AsH_3	Ozonizer	As_2H_4
B_2H_6	Ozonizer; Glow discharge between Cu electrodes	B_4H_{10}, B_5H_9, B_5H_{11}, B_6H_{10}, B_9H_{15}
B_5H_9 (+ H_2)	Glow discharge between Cu electrodes	$B_{10}H_{16}$ + ...
B_5H_9 (+ B_2H_6 + H_2)	Glow discharge between Cu electrodes	B_8H_{12} + ...
$B_{10}H_{14}$ (+ H_2)	Glow discharge between Cu electrodes	$B_{20}H_{16}$ + ...

Source: W. L. Jolly, in "Inorganic Synthesis with Electric Discharges," *Chemical Reactions in Electrical Discharges,* B. D. Blaustein, ed., Advances in Chemistry Series No. 80; American Chemical Society: Washington, D.C., 1969, p. 157.

References

1. H. Schüler and E. Lutz. *Z. Naturforsch.*, *12a*, 334 (1957).
2. R. L. McCarthy. *J. Chem. Phys.*, *22*, 1360 (1954).
3. H. Suhr. In *Techniques and Applications of Plasma Chemistry*, ed. J. R. Hollahan and A. T. Bell. Wiley: New York (1974).
4. M. A. Bondi. *Advan. Electron . . .* , *18*, 67 (1963).
5. W. H. Kassner and M. A. Bondi. *Phys. Rev.*, *137*, A317 (1965).
6. W. L. Fite. In *Chemical Reactions in Electrical Discharges*, ed. B. D. Blaustein. Advances in Chemistry Ser., *No. 80*. Amer. Chem. Soc.: Washington, D.C. (1969).
7. F. Kaufman. In *Chemical Reactions in Electrical Discharges*, ed. B. D. Blaustein. Advances in Chemistry Ser., *No. 80*. Amer. Chem. Soc.: Washington, D.C., 29 (1969).
8. E. E. Ferguson, F. C. Fehsenfeld, and A. L. Schmeltekopf. In *Chemical Reactions in Electrical Discharges*, ed. B. D. Blaustein. Advances in Chemistry Ser., *No. 80*. Amer. Chem. Soc.: Washington, D.C., 83 (1969).
9. A. G. Engelhardt and A. V. Phelps. *Phys. Rev.*, *131*, A2115 (1963).
10. A. G. Engelhardt, A. V. Phelps, and C. G. Risk. ·*Phys. Rev.*, *135*, A1566 (1964).
11. C. C. Goodyear and A. Von Engel. *Proc. Phys. Soc.*, *79*, 732 (1962).
12. A. Guntherschulze. *Z. Phys.*, *42*, 673 (1927).
13. R. W. Lunt and A. H. Gregg. *Trans. Far. Soc.*, *36*, 1062 (1940).
14. J. B. Thompson. *Proc. Roy. Soc.* (London), *A262*, 519 (1961).
15. J. B. Thompson. *Proc. Roy. Phys. Soc.*, *73*, 818 (1959).
16. J. W. Dettmer. Ph.D. Dissertation, Air Force Inst. of Tech. (1978).
17. F. Kaufman. In *Chemical Reactions in Electrical Discharges*, ed. B. D. Blaustein. Advances in Chemistry Ser., *No. 80*. Amer. Chem. Soc.: Washington, D.C. (1969).
18. F. C. Fehsenfeld, E. E. Ferguson, and A. L. Schmeltekopf. *J. Chem. Phys.*, *44*, 1844 (1966).
19. R. A. Young, C. R. Gatz, and R. L. Sharpless. *J. Phys. Chem.*, *69*, 1763 (1964).
20. R. N. Varney. *J. Chem. Phys.*, *13*, 1314 (1958); *33*, 1709 (1960).
21. C. F. Griese and W. B. Maier. *J. Chem. Phys.*, *35*, 1913 (1961).
22. J. F. Greenblatt and C. A. Winkler. *Can. J. Res.*, *27B*, 721 (1949).
23. B. J. Fontana. *J. Appl. Phys.*, *29*, 1688 (1958).
24. R. A. Young, R. L. Sharpless, and R. Stringham. *J. Chem. Phys.*, *40*, 111 (1964).
25. E. Rabinowitch. *Trans. Far. Soc.*, *23*, 283 (1937).
26. L. T. Herron, J. L. Franklin, P. Brandt, and V. H. Dibeler. *J. Chem. Phys.*, *30*, 879 (1958).
27. R. A. Young and G. A. St. John. In *Chemical Reactions in Electrical Discharges*, ed. B. D. Blaustein. Advances in Chemistry Ser., *No. 80*. Amer. Chem. Soc.: Washington, D.C. (1969).
28. B. D. Blaustein and Y. C. Fu. In *Chemical Reactions in Electrical Discharges*, ed. B. D. Blaustein. Advances in Chemistry Ser., *No. 80*. Amer. Chem. Soc.: Washington, D.C. (1969).
29. P. Cappezzuto, F. Cramarossa, P. Maione, and E. Molinari. *Grazz. Chim. Ital.*, *104*, 1109 (1974).
30. R. L. McCarthy. *J. Chem. Phys.*, *22*, 1360 (1954).

31. P. Cappezzuto, F. Cramarossa, G. Ferraro, P. Maione, and E. Molinari. *Grazz. Chim. Ital.*, *103*, 75 (1973).

32. J. L. Weininger. *J. Phys. Chem.*, *65*, 941 (1961).

33. A. N. Wright and C. A. Winkler. In *Active Nitrogen*. Academic Press: New York (1968).

34. M. Hudies and T. Wydeven. *J. Polym. Sci., Polym. Letters Ed.*, *13*, 549 (1975).

35. J. R. Hollahan and B. B. Stafford. *J. Appl. Polym. Sci.*, *13*, 807 (1969).

36. J. R. Hollahan and R. P. McKeever. In *Chemical Reactions in Electrical Discharges*, ed. B. D. Blaustein. Advances in Chemistry Ser., *No. 80*. Amer. Chem. Soc.: Washington, D.C. (1969).

37. C. Ponnamperuma, F. Woeller, J. Flores, M. Romiez, and W. Allen. In *Chemical Reactions in Electrical Discharges*, ed. B. D. Blaustein. Advances in Chemistry Ser., *No. 80*. Amer. Chem. Soc.: Washington, D.C. (1969).

38. A. Brown. *Compt. Rend.*, *152*, 1850 (1911).

39. ICI, Ltd. British Pat. 966,406 (1964).

40. E. A. Ogryzlo. *Can. J. Chem.*, *39*, 2556 (1961).

41. G. C. Fettis and J. H. Knox. In *Progress in Reaction Kinetics*, Vol. 2, ed. G. Porter. Pergamon Press: Oxford (1964).

42. B. A. Thrush. In *Progress in Reaction Kinetics*. Vol. 3. Pergamon Press: Oxford (1965).

43. H. G. Wagner and J. Wolfrum. *Angew. Chem.*, *83*, 561 (1971).

44. K. Hoyermann, H. G. Wagner, J. Wolfrum, and R. Zellner. *Chem. Ber.*, *75*, 22 (1971).

45. R. Klein and M. D. Scheer. *J. Am. Chem. Soc.*, *80*, 1007 (1958).

46. W. H. Stueben and J. Heicklen. *J. Chem. Phys.*, *46*, 4843 (1967).

47. T. W. Shannon and A. G. Harrison. *Can. J. Chem.*, *41*, 2455 (1963).

48. F. K. McTaggart. In *Plasma Chemistry in Electrical Discharges*. Elsevier: Amsterdam (1967).

49. S. M. L. Hamblyn and B. G. Reuben. *Adv. Inorg. Chem. Radiochem.*, *17*(J), 89 (1975).

50. H. Suhr and R. I. Weiss. *Z. Naturforsch.*, *256*, 41 (1971).

51. H. Suhr and G. Kruppa. *Liebigs Ann. Chem.*, *1*, 744 (1971).

52. H. Suhr. Chem. Inst. of Univ. of Tübingen, Germany (1973).

53. J. R. Hollahan and R. P. McKeever. In *Chemical Reactions in Electrical Discharges*, ed. B. D. Blaustein. Advances in Chemistry Ser., *No. 80*. Amer. Chem. Soc.: Washington, D.C. (1969).

54. H. Suhr. *Z. Naturforsch.*, *23b*, 1559 (1968).

55. H. Suhr, U. Schöch, and U. Schücker. *Synthesis*, 426 (1971).

56. H. Suhr and A. Szabo. *Liebigs Ann. Chem.*, *37*, 752 (1971).

57. H. Suhr and G. Rosskamp. *Liebigs Ann. Chem.*, *43*, 742 (1970).

58. S. D. Gokhale and W. L. Jolly. *Inorg. Chem.*, *3*, 1141 (1964).

59. R. N. Grimes. *J. Amer. Chem. Soc.*, *88*, 1895 (1966).

60. W. L. Jolly. In *Chemical Reactions in Electrical Discharges*, ed. B. D. Blaustein. Advances in Chemistry Ser., *No. 80*. Amer. Chem. Soc.: Washington, D.C. (1969).

4. Plasma Polymerization

1. Coupling of Electrical Energy into Plasma

In this chapter the conditions and results of polymerization under a variety of plasma conditions are described. The numerous scientific data reported in the literature have been obtained under conditions that differ not only in terms of conventional parameters but also in their mode of coupling. Therefore, it appears convenient to review this aspect of plasma polymerization briefly.

Almost all types of electrical discharges discussed in this book have three elements in common: (a) they are sustained by a source of electrical power; (b) this power is delivered by means of a coupling mechanism; and (c) it is delivered to a plasma environment associated with the particular device (see table below). Perhaps the most meaningful of the three above-mentioned elements in terms of classification is the way in which electrical energy is coupled into a given plasma environment [1].

Power source	Coupling mechanism	Plasma environment
d.c.	resistive	current level
a.c.	capacitive	pressure
r.f.	inductive	gas flow, temperature, etc.
Microwave		boundaries
		solid, liquid phases
		electrical fields
		magnetic fields

In those systems that have electrodes in direct contact with the ionized gas or glow discharge, the mode of coupling is resistive. Here, the electrical field necessary to sustain the plasma is caused by positive and negative charge accumulations both within and at the boundaries (walls or electrodes) of the plasma region. A finite potential difference at the gas-electrode boundary always exists because of the accumulated charges. The potential supports the various collision processes (ionization, excitation, electron emission or collection, and so forth) that operate to sustain the discharge.

On the other hand, inductively or capacitively coupled discharges do not have electrodes in direct contact with the gaseous plasma and are therefore also referred to as "electrodeless" discharges. The electric field is generated or "induced" by the inductively coupled discharge from a changing magnetic field. In contrast to resistively coupled discharges where electrons flowing in the external circuitry enter or leave at the electrode boundaries, in the inductively or capacitively coupled discharges, electrons are generated entirely within the discharge device, causing current to flow in closed loops. In analogy to a transformer, the plasma environment acts as a one-turn (short-circuited) secondary coil.

In capacitively coupled discharges, also referred to as polarization discharges, the electrodes through which the electrical power is delivered are physically separated from the plasma region by a nonconductive wall, that is, a dielectric barrier, which is in direct contact with both the electrodes on one side and the plasma region on the other.

In contrast to inductive discharges, the electrical field in the plasma region is caused by oscillating electrostatic charges at the dielectric barrier surfaces rather than by a changing magnetic field in the plasma region. In order to obtain such coupling, a high-voltage oscillating power supply is required to permit a displacement current to pass through the barrier material [2]. Discharges are also known that have a combination of these coupling mechanisms [3].

A plasma environment can be generated for a given power source, mode of coupling, and device geometry. The properties of this environment can readily be altered by manipulation of the externally controlled variables, such as gas flow rate, direction, inlet temperature, frequency, current, voltage level, duration of input power, presence or absence of a magnetic field, etc. It is evident that manipulation of any of these variables will affect chemical reactions in the plasma.

Since there exists insufficient knowledge as to the true nature of chemical reactions involving ions, electrons, radicals, photons, and so forth, very little is known or reported concerning the true state of plasma environment. Therefore, meaningful comparisons of experimental results for the same reaction in different devices are generally difficult to conduct. It is known, however, that the same reaction can be carried out in discharges that differ greatly with respect to the state of the plasma environment.

A large number of polymerization data provided by the literature were obtained from glow discharges between electrodes within the reaction vessel also referred to as "resistive coupling." When d.c. discharge is employed in such a system, say at pressures of about 0.5 to 5.0 torr, electrons originating near the cathode experience the electrostatic repulsion of the cathode and move into the Crooke's dark space (chapter 2, figure 2.5), which is the region of the greatest potential drop. Here, the electrons are accelerated to their maximum energy, which depends on the potential drop and the pressure of the discharge. The electrons emerge from the Crooke's dark space and enter the negative glow, where they bombard the gas molecules and thereby generate the wide range of gaseous species discussed in the previous section. Positive ions formed in the negative glow move into the Crooke's dark space and are accelerated by the high field in this region. The ions then impact upon the cathode, liberating more electrons, which repeat the above sequence. The electrons leaving the negative glow have lost most of their energy in collisions with gas molecules.

When alternating current is employed instead of direct current, then for the purpose of this discussion, each electrode may be regarded as acting alternately as the cathode. The description given above and in previous sections is general for all glow discharges. For an a.c. glow discharge occurring in an organic vapor, polymer coatings of equal thickness will form on the electrode surfaces.

It has generally been observed that the rate of polymerization is closely related to the degree of monomer adsorption on the electrode surface and that under normal discharge conditions, polymerization occurs at that surface. Thus, the electrode surface becomes the polymer that is being deposited at the electrode.

At frequencies above 1 MHz, direct contact between the electrodes and the plasma is no longer necessary. The energy can be fed to the plasma indirectly by capacitive or inductive coupling. Since there is little frequency dependence of plasma reactions, the choice of operating frequency is dictated by aspects of handling, shielding, and availability. The various kinds of electrical discharges have been found to lead to nearly identical results [4]. Therefore, except for noting the mode of coupling, polymerization results are not treated separately here.

2. Mechanistic Aspects

A. Nature

In the literature, distinction is frequently not made between "polymerization mechanism" and "polymer deposition mechanism." Also, "polymer deposition rate" is often treated as synonymous with "polymerization rate." It should be noted that in reactions in which the monomer is not completely converted to polymer, the distribution of polymer deposition cannot be related to an absolute polymerization rate. Other factors, discussed later in this chapter, also need to be considered. Furthermore, polymerization can initiate and take place both in the gas phase and on the surfaces of the electrodes or reactor walls. While these aspects will not be treated in a separate section, such distinctions will be made throughout the following discussions.

In conventional gas-phase polymerizations by means of photochemical or free-radical initiation, only vinyl compounds can be polymerized, and unless the monomer contains polar groups, only low-molecular-weight products are obtained. Furthermore, such monomers as styrene, acrylonitrile, methyl isopropenol ketone, and vinyl chloride do not polymerize in the gas phase unless they are photosensitized. Generally, monomer vapor pressures for these polymerizations on the order of 30 torr or higher are required.

In contrast, plasma-initiated gas-phase polymerization is not limited to the vinyl class of monomers. Many different types of monomers can be polymerized, including those that are not polymerized by any of the conventional polymerization techniques. In general, insoluble material is obtained and in most cases deposited or formed on the surface as a film. Under certain glow-discharge conditions, however, soluble materials of lower molecular weight may be obtained or, under other conditions, polymer material in the form of powder may result.

The conversion of volatile organic compounds into liquid and solid products by the action of a high-voltage gas discharge has been known for over a century [5]. It was not until the early sixties, however, that some resources were redirected from pure research to product application, such as coatings for containers, steel strips, and fabrics. Hay [6] reported in 1969 on the polymerization of vinyl monomers in an electrical discharge at atmospheric pressure. He expected that the monomer ionization would initiate a polymerization reaction by an ionic mechanism or that an organic ion would decompose to a free radical and start a free-radical polymerization in a nitrogen atmosphere.

The compounds that gave the best polymer yields included triallyl amine, acrylonitrile, toluene, and styrene. Hay suggested that the initiation step proceeds via cation formation:

$$N_2 + e^- \rightarrow N_2^+ + 2e \tag{1}$$

or

$$RH + e^- \rightarrow RH^+ + 2e \tag{2}$$

Considerable attention has been given by researchers to the plasma polymerization of low-molecular-weight hydrocarbons, and it was found that compositions of the polymers obtained from the excitation of hydrocarbon vapors are as varied as their modes of formation. It is interesting to observe that at the temperature of liquid air, the polymerization of methane in a glow discharge was reported [7] to produce polymethylene, $(CH_2)_n$.

Bell and coworkers [8] have studied the effects of reaction conditions on the plasma polymerization of ethylene and have found that the form of the resulting polymer may be a powder at low pressure (1–2 torr) and low monomer feed rate (10–40 cc/min), or a colorless film at low pressure and high feed rate (70–90 cc/min), or an oil at higher pressure (4–5 torr) and high feed rate. Both the powder and film are insoluble in common organic solvents, indicating a high degree of crosslinking. The oily products were soluble in both acetone and xylene, however.

It will be shown in the subsequent section that the hydrocarbon oil consists of highly branched oligomers containing primary, secondary, and tertiary carbon atoms as well as small amounts of olefinic unsaturation and pendant aromatic groups. Because of the complexity of the polymerization kinetics involved in plasma polymerization and the lack of basic data, a quantitative interpretation of the mechanisms has not yet been possible. Bell and his coworkers have shown [9,10], however, that a simpler plasma reaction can be interpreted in terms of the physical characteristics of the plasma.

The main energy carriers in a glow discharge are the free electrons generated by a partial ionization of the gas carrier. They are characterized by a distribution of energies and the average energy of a few electron volts. The population of electrons, having higher than average energies, decreases rapidly with increasing energy. But there will be a small number of electrons that may possess energies of as much as 10 to 20 eV.

A variety of species are generated by collisions between the free electrons and the monomer; they include excited molecules, free radicals, and ions. The concentration of free radicals has been found to be five to six orders of magnitude larger than that of the ions; therefore they have been considered [8] to be the primary precursors to stable products.

In analogy with the photolysis of ethylene [11] and the polymerization of ethylene by gamma radiation [12] it has been proposed [8] that the process of plasma polymerization follows a free-radical mechanism:

$$e + C_2H_4 \rightarrow C_2H_2 + H_2 + e \tag{3}$$

$$e + C_2H_4 \rightarrow C_2H_2 + 2H\cdot + e \tag{4}$$

These two reactions are in agreement with large amounts of acetylene found by several investigators in the gaseous products.

The hydrogen atoms from reaction 4 can react with other species in the plasma to produce additional radicals and can thus initiate a gas-phase polymerization. It has been suggested [13] that the powder polymer obtained during vigorous polymerization of ethylene in plasma is the result of such a free-radical process. Growth termination can occur at any stage by combination with other radicals or with an atom of hydrogen.

At the low pressures generally employed in glow-discharge polymerizations, a fraction of the free radicals generated in the gas phase can be expected to diffuse to the reactor walls, at which a number of reactions can take place, such as those described by Bell and coworkers [8]:

$$M + S \rightleftharpoons S{-}M \tag{5}$$

$$R_{\dot{n}} + S{-}M \rightarrow S{-}R_{\dot{n}+1} \tag{6}$$

$$R_{\dot{n}} + S \rightarrow S{-}R_{\dot{n}} \tag{7}$$

$$M + S{-}R_{\dot{n}} \rightarrow S{-}R_{\dot{n}+1} \tag{8}$$

$$S{-}M + S{-}R_{\dot{n}} \rightarrow S{-}R_{\dot{n}+1} + S \tag{9}$$

$$S{-}R_{\dot{m}} + S{-}R_{\dot{n}} \rightarrow S{-}P_{m+n} \tag{10}$$

$$R_{\dot{m}} + S{-}P_n \rightarrow S{-}P_{m+n} \tag{11}$$

where M is a monomer molecule (ethylene or acetylene), R$\cdot$ is a free radical of n units, S is a surface site, and P_n is a terminated polymer chain.

Other radicals may also be generated by collision of energetic ions and absorption of UV radiation from the plasma. The observed [8] H/C ratios of 1.3 to 1.5 can derive either from reactions of acetylene with ethylene or from hydrogen removal from the polymer by UV absorption or collisions with energetic species.

The dependence of both homogeneous and heterogeneous polymerization processes on free radicals as initiators permits the postulation that the overall rate of polymerization is related to the gas-phase concentration of such free radicals. Since the radicals are formed by reactions requiring atomic hydrogen, high concentrations of free radicals will correlate with high concentration of atomic hydrogen. The concentration of this latter species is dependent upon the rate of its formation from reaction 4 and the rate of its disappearance by convective transport, reaction, and recombination.

These researchers [8] expressed the rate of atomic hydrogen formation according to reaction 4 by

$$r_2 = k_2[e] \, [C_2H_4] \tag{Eq. 4.1}$$

where $[e]$ and $[C_2H_4]$ are the concentrations of electrons and ethylene, respectively. The rate constant, k_2, is dependent on the average electron energy,

E/p, that is, the ratio of the electric field strength, needed to sustain the discharge, to the gas pressure [11]. Brown has earlier shown [14] that E/p is related to $p\Lambda$, where $\Lambda = d/\pi$ and d is the electrode separation. If the $\bar{P}$ is the power density, the quantity $[e]/Pd$ can also be expressed as a function of $p\Lambda$. Therefore, k_2 and $[e]$ can be determined from p, d, and $\bar{P}$.

Bell and his coworkers have shown [9,10] that the aforementioned relationships between k_2 and E/p and between E/p and $p\Lambda$ can also be used to describe the dissociation of hydrogen and oxygen, the oxidation of carbon monoxide, and the decomposition of carbon dioxide.

It has been reported [8] that the overall rate of plasma polymerization of ethylene goes through a maximum as the monomer flow rate increases from 5 to 80 cm^3/min. At low flow rates a steady-state free-radical concentration can be assumed to be independent of the gas flow rate. Thus, initiation steps for both homogeneous and heterogeneous polymerization by free radicals are likely to take place. As the flow rate is raised, the larger supply of monomer will initially increase the polymerization rate until a point is reached at which the flow rate causes a reduction in the free-radical concentration and consequently a decrease in the rate of polymerization.

Because of the numerous parameters involved, the influence of pressure on the polymerization rate is considered to be most difficult to assess. With rising pressure, E/p decreases, and therefore the value of k_2 is reduced, together with the magnitude of $[e]$. While this reduction in k_2 and $[e]$ operates to decrease the rate of H formation, it is partially compensated for by the rise in ethylene concentration.

By analogy with other systems [9,10] it has been suggested [8] that the rate of atomic hydrogen formation, r_2, will increase with pressure and will thus cause the free-radical production rate to rise. This trend is opposed by processes that consume free radicals and are also enhanced by increases in the pressure of ethylene. Thus, Bell and his coworkers [8] suggest that as a result of these two processes, the free-radical concentration will increase with pressure but at a decreasing rate. Furthermore, the concentration of monomer present on the surfaces of the deposited film will also increase with rising ethylene pressure until a maximum concentration of adsorbed monomer equals a monolayer coverage.

On the basis of the foregoing arguments, the investigators explain the observed dependence of the rates of film and powder formation on pressure in terms of the anticipated responses of gas-phase free-radical concentration and the surface concentrations of adsorbed monomer, as follows.

As the pressure is increased from very low values, the free-radical concentration, the monomer concentration, and the concentration of adsorbed monomer all increase. Over this pressure range the rates of both film and powder formation will rise. When the pressure is increased further, gas-phase polymerization will decline because of chain termination reactions that are second order in free-radical concentration. At the same time, the rate of heterogeneous polymerization will approach a maximum as a result of saturation of the surface with adsorbed monomer. It is evident from the preceding discussions that the use of a glow discharge to initiate polymerization is a fairly violent method and can lead to fragmentation of the original monomer, with the resultant incorporation of such fragments into the polymer. At sufficiently high frequencies, such as are generally employed in most laboratories, positive ions will not be able to follow the field and will move only by diffusion. Energy is transferred to the neutral molecules by inelastic collisions with electrons to result in excited or ionized species. Recombination of positive ions and electrons gives rise to highly excited states, which may dissociate into fragments believed to be predominantly free radicals.

While it cannot be said conclusively whether polymerization is initiated exclusively by ions or radicals, most researchers today tend to favor the latter and the convincing arguments mentioned above. The presence of free radicals in plasma-deposited films has been demonstrated in numerous laboratories directly by measurements, indirectly, and by exposure of deposited films followed by exposure to air, after which the extent of postoxidation is measured.

As an example, it has been shown [15] that films obtained from low-current-density polymerization of styrene are only slightly crosslinked, while at a high current density of over 3 mAcm^{-2}, highly crosslinked solid particles were obtained. The observation of interest here is that the polymer films obtained at low current densities were tacky and readily soluble in benzene. The presence of trapped radicals in the films was established by reaction with 1,1-diphenyl-2-picrylhydrazyl (DPPH) in solution. From decay plots of DPPH the quantity of free radicals trapped during the glow-discharge polymerization was estimated to be one trapped free radical for every 500 monomer units in the film.

On exposure of the polymer film to air, the number of free radicals decreased rapidly with time. No radicals were detected after 48 hr exposure to air. At the same time, the tacky film became increasingly harder, also manifesting a weight increase and a decrease in solubility. These changes suggested that oxygen is reacting with trapped free radicals, thus leading to both increased molecular weights and crosslinking.

It is well documented that irradiation of solid polymers produces trapped radicals that can subsequently be used in other reactions, such as grafting. It has also been shown [16] that exposure of a polymer to an electric discharge can produce radicals trapped in the polymer. Radicals have also been created by the bombardment of adsorbed monomer molecules by electrons. They can react in a propagation step with monomers.

$$R_{\dot{n}} + M \rightarrow R_{\dot{n}+1} \tag{12}$$

Radicals can also combine or disproportionate in a bimolecular process, and they may become trapped in a unimolecular process.

$$R_{\dot{n}} + R_{\dot{m}} \rightarrow P_{n+m} \ (\text{or } P_n + P_m) \tag{13}$$

$$R_{\dot{n}} \rightarrow R_{\dot{n}} \ (\text{trapped}) \tag{14}$$

If the rate of production of free radicals is r, the rate of deposition of polymer is given by

$$\text{Rate} = r \cdot \frac{v_1}{v_1 + v_2 + v_3} \tag{Eq. 4.2}$$

where v_1, v_2, and v_3 are the velocities of reactions 12, 13, and 14, respectively. Expressing these reaction velocities in terms of the specific rate constants and the concentrations of reactants,

$$\text{Rate} = r \ \frac{k_1[R\cdot][M]}{k_1[R\cdot][M] + k_2[R\cdot]^2 + k_3[R\cdot]} \cdot \tag{Eq. 4.3}$$

In order to express [M] in terms of monomer pressure in the system, use can be made of the evidence [17] that the reaction of adsorbed monomer is involved in the polymerization process. If a radical is formed on the surface adjacent to an adsorbed monomer molecule, the rate of reaction must be proportional to the amount of monomer adsorbed.

The adsorption behavior of the monomer has been represented by the equation for multilayer adsorption:

$$\frac{p}{v(p_0 - p)} = \frac{1}{v_m C} + \frac{(C-1)p}{v_m C} \cdot \frac{p}{p_0} \tag{Eq. 4.4}$$

where v is the volume of gas adsorbed, v_m is the volume of a monolayer, p is the pressure in the system, p_0 is the saturated vapor pressure of the adsorbate at the experiment temperature, and C is a constant. Equation 4.4 has generally been found to be in good agreement with experimental observations up to relative pressures (p/p_0) of about 0.4.

The shape of the curve from equation 4.4 is chiefly determined by the magnitude of C, which is given approximately by [18]

$$C \simeq \exp(\Delta H_L - \Delta H_1)/RT \tag{Eq. 4.5}$$

where ΔH_L is the heat of liquification of the adsorbate and ΔH_1 is the heat of adsorption of the first molecular layer. It can be seen from equation 4.4 that with $C = 1$, the volume of gas adsorbed is not equal to the monolayer volume until the relative pressure is 0.5 torr.

In many cases the surface of the electrode will not be completely covered, and a radical site may be formed at a point on the surface where no adsorbed monomer is adjacent. Such a radical can participate in a propagation reaction only when a monomer diffuses from the gas phase to the site. The rate of monomer arrival at the surface will be proportional to its pressure.

On the basis of these arguments, the rate of the propagation step will be proportional to monomer pressure regardless of whether a radical reacts with adsorbed monomer or with monomer diffusing from the gas phase. This would give the simplified relations:

$$[M] = kp \qquad \text{(Eq. 4.6)}$$

where p is the pressure of monomer in the system. Substituting in equation 4.3 and simplifying,

$$\text{Rate} = r \left\{ p / \left(p + \frac{k_2}{kk_1} [R \cdot] + \frac{k_3}{kk_1} \right) \right\} \qquad \text{(Eq. 4.7)}$$

Equation 4.7 indicates that at high monomer pressures the rate will tend to a limiting value of r.

For a given set of conditions in the r.f. polymerization of styrene between two electrodes [19] in the pressure range of 0.5 to 3.0 torr, the limiting rates were found to be dependent on the power, as shown in table 4.1. These limiting rates are equal to the rates of radical production within the power levels shown in table 4.1. If radicals are generated by excitation upon electron impingement followed by dissociation,

$$P \rightarrow P^* \rightarrow 2R^* \qquad (15)$$

the rate of radical formation would be expected to be linearly dependent upon power. Alternatively, if radicals are produced by dissociation of the highly excited states formed by recombination of ions and electrons,

$$P^+ + e \rightarrow P^{**} \rightarrow 2R \cdot \qquad (16)$$

the rate of radical generation should be dependent on the square of the power.

Under the aforementioned conditions in the polymerization of styrene [18], a plot of log r versus log W indicates that r in these experiments is propor-

Table 4.1. Limiting rates at higher pressures

Power (W cm^{-2})	Limiting rate (μg cm^{-2} sec^{1})
0.4	1.8
1.5	17.5
2.7	44.4

Source: A. R. Denary, P. A. Owens, and A. Crawshaw, *Europ. Polym. J., 4,* 93 (1968). Reprinted with permission from *European Polymer Journal.* Copyright 1968, Pergamon Press, Ltd.

tional to $W^{1.7}$, which suggests that radicals are being generated by both first- and second-order processes in power, that is:

$$r = k''W^2 + k'W \qquad \text{(Eq. 4.8)}$$

or

$$r/W = k''W + k' \qquad \text{(Eq. 4.9)}$$

In order to obtain a value for [R·] in equation 4.7, the stationary-state treatment may be applied to the radicals:

$$\frac{d[\text{R}\cdot]}{dt} = r - k_2 [\text{R}\cdot]^2 - k_3 [\text{R}\cdot] = 0 \qquad \text{(Eq. 4.10)}$$

Polymerization reactions in glow discharges may take place in both the gas phase and at the electrode or wall surface. Since the rate of polymerization is closely related to the degree of monomer adsorption, it is probable that polymerization occurs chiefly at the solid surface. It will be seen later that at higher pressures, the oligomers can also be formed in the gas phase.

The formation of the active species, which initiate polymerization reactions, may occur either in the gas phase or on the electrode or wall surface. It is known that electron bombardment of organic vapors in the negative glow of a discharge produces a wide range of species, many of which could initiate polymerization reactions by condensing on the solid surface and reacting with the unactivated adsorbed molecules. There is much evidence that this type of reaction does not play a major part, however. It has generally been found that many polymers in discharges exposed to gases such as oxygen, nitrogen, hydrogen, and argon produce radicals characteristic of the polymer formed on the surface. These discharge-induced surface radicals are similar to those obtained by irradiation with X-rays or high-energy electrons.

The primary step when an ion impacts on a polymer surface is probably the same as that observed in high-energy radiation—excitation of the polymer molecules, followed by fragmentation yielding active species that can propagate the polymerization reaction. The adsorption of a monomer at constant temperatures on an electrode had earlier been expressed by the equation [17]:

$$x/m = -\int (p/p_0)T \qquad \text{(Eq. 4.11)}$$

where x is the amount adsorbed and p/p_0 is the relative vapor pressure, that is, the pressure after adsorption divided by the saturated vapor pressure at the operating temperature, thus representing the amount of monomer adsorbed on the surface.

The rates of polymerization of various monomers determined at room temperature and within 1–3 torr pressure have been correlated with the amount of monomer adsorbed on the electrode surface (p/p_0) in table 4.2.

Williams and Hayes [17] suggest that the amount of monomer adsorption on the surface of the electrode is extremely important in determining the rate of polymerization of such monomers in a glow discharge. The data in table 4.2 were obtained at about 10 Kc/sec with a glow sustained at 400 V in a discharge current varying between 1 and 5 m amp/cm^3 in hydrogen, nitrogen, or noble gases. Thus, in the polymerization of styrene in discharges between electrodes, it was postulated that polymerization occurs at the surface of the

Table 4.2. Rates of polymerization of various monomers in relation to adsorption

Rate of polymerization	Monomer	Approximate value of p/p_0
High	styrene	0.2
	α-methyl styrene	0.2
	dicyclopentadiene	0.2
	acrylic acid	0.2
Moderate to slow	allyl alcohol	0.05
	ethyl acrylate	0.03
	methyl methacrylate	0.03
	methyl acrylate	0.01
Slow	vinyl acetate	<0.01
	isoprene	0.002
	isobutene	0.001
	ethylene	0.001

Source: T. Williams and M. W. Hayes. Reprinted by permission from *Nature*, London, *209*, 769 (1966). Copyright © 1966 Macmillan Journals Limited.

electrode. The dependence of the polymerization rate on the wattage and the momomer pressure supported the mechanism of polymerization of adsorbed monomer on the electrode surface.

In electrodeless discharges, however, deposition of polymers occurs on the wall of the discharge vessel. Yasuda and Lamaze [19] concluded that under the experimental conditions of a styrene flow rate of about 6.0×10^{-2} cm³/ min at 0.02 mm Hg and r.f. power supply at 13.56 MHz, most polymerization is initiated and propagated in the vapor phase. As propagation proceeds, the growing polymer is deposited on the walls of the discharge vessel, since it cannot remain in the vapor phase. Under this assumption, the termination of the polymer chain takes place on the wall, and the deposition rate is directly proportional to the propagation rate of growing polymer in the gas phase. For these conditions, the investigators proposed the following kinetic scheme.

Excitation Step:

$$M \xrightarrow{k_1} M* \tag{17}$$

$$X \xrightarrow{k_2} X* \tag{18}$$

$$X* + M \xrightarrow{k_3} M* + X \tag{19}$$

Initiation Step:

$$M* \xrightarrow{k_i} M* \tag{20}$$

Propagation Step:

$$M \cdot + M \xrightarrow{k_p} MM \cdot \tag{21}$$

$$M \cdot_n + M \xrightarrow{k_p} M \cdot_{n+1} \tag{22}$$

where M is the monomer, M* is the excited monomer, X is the plasma gas, and X* is the excited plasma gas: Then the rate of propagation is given by [M] and [X] as follows:

$$\text{Rate of propagation} = k_i k_p k_1 [M]^2 + k_i k_p k_2 [M]^2 [X]$$
$$= k_i k_p k_1 [M]^2 \{1 + k_2 k_3 [X]/k_1\} \quad \text{(Eq. 4.12)}$$

H_2, N_2, He, and A were used as plasma gas in these experiments.

It is noted that the polymer deposited by H_2 plasma was clear and colorless, while polymers deposited in the presence of He, Ne, and A were light amber.

Thus, some hydrogenation is assumed to have occurred but was not found to affect the deposition rate to any extent. All polymers were apparently highly crosslinked, since they did not dissolve in any solvent.

When the monomer-vapor particle pressure in the chamber was maintained between 0.6 and 2.80 torr, powder formation was observed [19] from the following monomers tested: styrene, toluene, benzene, p-xylene, hexane, isoprene, acetonitrile, vinyl chloride, tetrabutyltin, styrene/divinylbenzene and styrene/1,2-dibromoethane.

It has been suggested that electron-impact ionization is the major source for charged-species generation in a glow discharge. Another important ionization mechanism is the direct ionization by collision with sufficiently energetic, metastable neutral species, however. This process is known as the Penning ionization and becomes the major mechanism whenever noble gases, such as He and Ne, are used as one of the gas components in the reactor [20]. For He, Ne, and A, the excitation energies of the lowest triplet state are 19.80, 16.62, and 11.55 eV, respectively. Although the lowest triplet state in "active nitrogen" does not exist because of the very efficient quenching of it by the nitrogen atom [21], "active nitrogen" is unique because of its extensive vibrational excitation which persists for a long time beyond the discharge zone and thus leads to extensive chemiionization [20]. The ionization potential for most neutral gases is in the neighborhood of 15 eV.

The amount of polystyrene powder obtained under the aforementioned conditions [22] decreased from He to N_2 to Ne to A to air. The data, except for air, suggest that the Penning ionization mechanism operated in the initiation of this polymerization. The dependence of yield of polystyrene powder on inert gas is shown in table 4.3. A particle size of about 1350 Å was found to be the most common diameter under the test conditions cited [22], while at pressures below 0.6 torr styrene was found to form a film.

It is intriguing to speculate as to the exact species that initiates this polymerization. The chemical reaction of active hydrogen in H_2 plasma could

Table 4.3. Dependence of yield of polystyrene powder on type of inert gas

Inert gas	Yield (%)
Helium	18
Nitrogen	16
Neon	15
Argon	7
Air	3

Source: R. Liepins and K. Sakaoku, J. Appl. Polym. Sci., 16, 2633 (1972). Copyright © 1972, John Wiley & Sons, Inc. Reprinted by permission of John Wiley & Sons, Inc.

be due to one or more of the following species: hydrogen atom H, excited hydrogen atoms H*, H_2^+ ions, excited hydrogen molecules H_2*, protons H^+, or unstable complex H_K*. The energy required for excitation of the H_2 to the repulsive $^3\Sigma\mu$ triplet state that results in dissociation into two normal H atoms is about 9 eV (207 Kcal/mole), and the ionization potential for the formation of H_2^+ ions is about 15.4 eV.

In a novel approach toward preparing plasma polymers, it was demonstrated more recently [23] that a low-pressure plasma can be used to initiate conventional polymerization of vinyl monomers in the liquid state. Methyl methacrylate was sealed in a thin-walled ampule after degassing and was inserted between a pair of parallel-plate electrodes, which were connected to a r.f generator operating at 13.56 MHz frequency. The cooled ampule was permitted to warm up until droplets of liquid monomer appeared. A glow discharge was then generated but self-quenched within 30–60 sec. Apparently, dissociation of the monomer gas through collisions with energetic electrons or ions increased the vapor pressure inside the ampule to such an extent that the discharge could no longer be sustained. Following initiation the ampule was left standing at constant temperature for defined time periods before breaking the seal. The rate of polymerization was estimated to be 4×10^{-6} mole/liter-sec. It is noted that comparable rates were obtained by gamma-ray polymerization using a dose rate of 2×10^{-2} rad/sec at 20°C [24], by beta-ray polymerization at a dose rate of 1.5×10^5 rep at 30°C [25], and by high-energy electrons at 1.5×10^5 rep total dose at 20°C [26]. In the first two cases, however, initiation was sustained throughout the polymerization time.

It is of interest to note that plasma-initiated polymerization of liquid monomers yields products of very high molecular weight in spite of the relatively high rate of polymerization. As an example, 10 min after initiation, polymethylmethacrylate had reached a molecular weight of $3 \cdot 10^6$ g/mole and 2 hours later 10^7 g/mole. By contrast, comparable cases of polymerization by γ-irradiation, β-rays, and high-energy electrons resulted in polymers of molecular weights of only about 5×10^5, 1×10^6, and 2×10^6 g/mole, respectively. The distribution of tacticity of 7% isotactic, 27% heterotactic, and 66% syndiotactic triads is characteristic of polymethyl methacrylate synthesized by conventional free-radical polymerization [27]. These data seem to support the theory that initiation of the above reaction by glow discharge occurs via a free-radical mechanism [4]. The investigators [22] polymerized by such plasma initiation, in addition to methyl methacrylate, other monomers, such as methacrylic acid, acrylic acid, methacrylamide, acrylamide, and several methacrylic acid esters, and they also copolymerized several of these monomer species. The unusual feature in this process is that initiation occurs in the gas phase, while propagation and termination are continued in the liquid.

B. Kinetics

The distribution of the electron velocity plays an important role in defining the physical properties of a plasma. Bell [28] derives from this:

The electron energy distribution.
The average electron energy.
The electron transport properties.
The rate constants for reactions involving electron-molecule collisions.

He has demonstrated that the shape of the distribution function depends upon the magnitude of the applied electric field and the nature of the elastic and inelastic interactions by the electrons.

The electron velocity distribution function, f, is described as representing the density of electrons in both position and velocity space. If the vector $\mathbf{r}$ describes a given point in position space and the vector $\mathbf{v}$ a given point in velocity space, the electron velocity distribution function f may be expressed as $f(\mathbf{r},\mathbf{v})$. The product of $f\,drdv$, in which $dr = dxdydz$ and $dv = dv_x dv_y dv_z$, defines the number of electrons whose positions are located within the volume element dr and whose velocities lie within the volume element dv.

The number of elements per unit volume, $f(dv)$, at r with velocities in dv is obtained from $f\,drdv/dr$. The electron density n at the point r is then obtained by integration of the product fdv; that is,

$$n = \int_{-\infty}^{+\infty} f\, dv \qquad \text{(Eq. 4.13)}$$

The single integral sign is used to denote a triple integration over the three coordinates in space.

Similarly if ϕ $(\mathbf{r},\mathbf{v})$ denotes some property of the electrons, which may be dependent on both position and velocity, the average value of this property weighted against the distribution of velocities can be described as

$$n\phi = \int_{-\infty}^{+\infty} fdv \qquad \text{(Eq. 4.14)}$$

Equation 4.14 can be used to determine the average electron energy. The velocity v can be expressed as $v = V + (v)$, where V is the random velocity and (v) the translational velocity of the whole electron cloud. The translational velocity, if for instance $\phi = v_x$, is then

$$(v_x) = \frac{1}{n} \int_{-\infty}^{+\infty} v_x\, fdv \qquad \text{(Eq. 4.15)}$$

Since positive and negative values of the random velocity, V, have equal probability, the average value of V will equal zero. The total translational kinetic energy, ϵ, can then be expressed by:

$$\epsilon = \frac{1}{2} m(v^2) = \frac{1}{n} \int_{-\infty}^{+\infty} \frac{1}{2} mv \ vfdv \qquad \text{(Eq. 4.16)}$$

which eventually becomes

$$\epsilon = \frac{1}{2}m(V^2) + \frac{1}{2} \ m(v)^2 \qquad \text{(Eq. 4.17)}$$

where v and V are the magnitudes of the velocities v and V, respectively.

The first term in equation 4.17 represents the kinetic energy associated with the random motion of the electrons, and the second term represents the kinetic energy associated with the translation of the electron cloud as a whole. If the velocity distribution is Maxwellian [8], the random kinetic energy can be related to the electron temperature, T_e, by

$$\frac{1}{2}m(V^2) = \frac{3}{2} \ kT_e \qquad \text{(Eq. 4.18)}$$

(As J. C. Maxwell noted in 1876, all possible velocities are present in a system with the existence of a most probable velocity, the probabilities decreasing with both lower and higher velocities.)

Bell [28] derives the exact form of the velocity distribution function from consideration of the gain and loss of electrons from an incremental volume in phase space defined by $drdv$. The net rate of transfer of electrons from this volume is summarized by the Boltzmann equation

$$\frac{\delta f}{\delta t} + v \cdot \nabla_r f + \frac{eE}{m} \cdot \nabla_v f = \left(\frac{\delta f}{\delta t} \right)_{\text{coll}} \qquad \text{(Eq. 4.19)}$$

In this equation are [28]:

$\dfrac{\delta f}{\delta t}$ = Local variation of the distribution function with time.

$\nabla_r f$ = Variation in the distribution function resulting from electrons streaming in and out of a given volume element (measure of diffusion).

$\dfrac{eE}{m} \nabla_v f$ = Variation of the distribution function resulting from an applied electrical field E acting on the electrons.

$\left(\dfrac{\delta f}{\delta t} \right)_{\text{coll}}$ = Net transfer of electrons from the differential volume by a mechanism of binary collisions between electrons and molecules, ions, and other electrons.

It has been mentioned in chapter 1 that many types of binary collisions can take place in a plasma. They can be conveniently divided into elastic and inelastic collisions. The elastic collisions involve electron impact with either neutral or charged targets in such a manner that no excitation of the target particle occurs. If the target is a molecule or ion, then there will take place only a small energy transfer from the electrons. But if interactions between two electrons result in a significant transfer of energy from one electron to another, the shape of the electron-velocity distribution function will be affected. In elastic collisions the nature of these excitations ranges across the energy spectrum from rotational excitations, in which 0.01 to 0.1 eV is absorbed, to ionizations in which more than 10 eV are absorbed.

In order to obtain the velocity distribution function, Bell [28], lacking an exact solution, sought an approximate solution by subdividing the distribution function f into an isotropic, f°, and anisotropic, $\phi(v)$, distribution; that is,

$$f = f^\circ + \phi(v) \qquad \text{(Eq. 4.20)}$$

By reasoning that the anisotropic contribution is largest when v is in the direction of the gradient (or force causing the perturbation) and is smallest when it is perpendicular to it, the anisotropic distribution is expressed by equation 4.21,

$$\phi(v) = \frac{v}{v} \cdot f' \qquad \text{(Eq. 4.21)}$$

in which f' points in the direction in which the electrons drift as the result of the external fields and spatial gradients. From equations 4.20 and 4.21 the total distribution is then:

$$f = f^\circ + \frac{v}{v} \cdot f' \qquad \text{(Eq. 4.22)}$$

Substituting equation 4.22, the approximate solution, into the Boltzmann equation results in the identification of two differential equations for f° and f'. This fairly complex procedure has been described by Allis [29].

Inelastic collisions contribute mainly to the distribution of velocities and relatively little to the transport properties associated with the electrons. The collision frequency, v_i, for each inelastic process is related to the cross section for that process by

$$v_i = Nv, \, \sigma_i(v) \qquad \text{(Eq. 4.23)}$$

Collisions arising from the interaction of electrons with either ions or other electrons become important only when a significant degree of ionization occurs [9]. In most high-frequency plasmas the degree of ionization in negligible, that is, less than 10^{-5}, to be considered in such collisions.

Further kinetic examination by Bell [28] led to the important conclusion that the average electron energy, T_e, is a function solely of E_0/p, the applied electric field divided by the pressure existing in the plasma. This is derived from the linear dependence of f_m on pressure, where f_m is the momentum transfer collision frequency. Representation of E_0/p as the average electron energy is confirmed by the proportionality of gas density and pressure.

C. Energy Transfer

The transfer of energy to a plasma takes place mainly through the action of the electric field on the electrons. The instantaneous rate of energy transfer per unit volume can be expressed by

$$p(t) = 6_r E_0 e^{i\omega t} \qquad \text{(Eq. 4.24)}$$

where 6 = conductivity and $E_0 e^{i\omega t}$ = alternating electric field. In order to arrive at the net power dissipation, equation 4.24 is averaged over the period of one cycle of the field [30].

D. Diffusion of Charged Species

At very low charge concentrations, that is, those observed near breakdown, electrons and positive ions will diffuse independently of each other. For steady-state conditions, the charged particles lost from the ionized gas by diffusion must be replaced by ionization. Bell [28] has demonstrated that at steady state, the process of free diffusion leads to a concentration of positive ions considerably higher than that of electrons. Under these conditions the ionized gas cannot be considered to be a plasma.

The separation of charge due to the difference in diffusion coefficients of electrons and ions generates a space-charge field, the magnitude of which will be small if the charge concentrations are low, $n_e < 10^2 \text{cm}^{-3}$. Under these conditions, the diffusion of either charge carrier will not be affected. As the charge concentration is increased to a level where the Debye length, λ_D, the distance over which a charge imbalance can exist, is much smaller than the diffusion length, Λ,

$$\lambda_D << \Lambda$$

then the medium is properly referred to as a plasma. Under these conditions, the space-charge field, E_{SC}, becomes sufficiently large to effect the transport of both electrons and positive ions. When a significant concentration of negative ions is present in addition to the positive ions, the diffusion of electrons becomes part of a three-particle system, that is, flux vectors for electrons, anions, and cations [31].

E. Wattage and Mass Flow Rate

It has been recognized that glow in plasma is dependent on flow rate and that deposition rate is greatly dependent on "glow character." Recently, a useful parameter for plasma polymerization has been introduced, namely, W/FM—energy input, or wattages per mass flow rate, where F is expressed in cm^3/min and M is the monomer molecular weight. It has been shown [28] that the chemistry involved in plasma polymerization is largely determined by the character of the glow and that comparison of two monomers should be made at comparable "glow character" instead of comparable flow, pressure, or discharge power. When the flow rate is increased at constant discharge power, the composite power parameter W/FM is therefore decreasing. When W/FM decreases beyond a threshold value, the glow character changes as manifested by polymer deposition rates. When deposition rates are compared at a fixed flow rate or at a fixed wattage, a monomer with the higher molecular weight is likely to show lower deposition rates when compared with a lower-molecular-weight monomer because the values of W/FM are smaller.

The applicability of W/FM as a composite power parameter to describe conditions of plasma polymerization has been demonstrated in several examples [28]. In discharge, the number of gas molecules in the system changes. While polymerization decreases the number of molecules in the gas phase, the splitting of a molecule (hydrogen formation, detachment of oxygen-containing groups) increases the total number of molecules. This change has been expressed by a parameter γ.

Thus, when the system pressure in a plasma is under varying wattage at various flow rates, it has been shown [28] that the plasma pressure for acetylene, $HC{\equiv}CH$, decreases with increasing power, while that for ethylene, $H_2C{=}CH_2$, increases with increasing power. In more general terms, for monomers that have low γ values, p_g decreases below the starting pressure p_0, while for monomers having high γ values, p_g increases beyond p_0. The value of p_0 at a fixed flow rate depends on the pumping rate. The relationships between p_g and γ and F have recently been described in detail [32]. But nevertheless, the plasma pressure, p_g, can be maintained at a constant value as the discharge power increases, and it remains constant at higher wattage. In the vicinity where p_g becomes constant, that is, independent of wattage, the characteristic glow changes from incomplete or weak glow to complete or full glow. At this critical wattage, W_c, the discharge that uses a higher wattage is characterized by full glow and the discharge at lower wattage by weak glow. W_c is dependent on flow rate F and molecular weight M and varies in the range from less than 10 S_c to over 150 S_c for many monomers. Values of W_c extrapolated to $F = 0$ are not much different for various monomers.

If the extrapolated value for W_c is W_0, then $(W_c - W_0)$ is referred to as the effective critical discharge power, W_e; that is,

$$W_e = W_c - W_0 \qquad \text{(Eq. 4.25)}$$

Typical plots of W_c versus F are illustrated in figure 4.1 for various monomers [32]. The effect of chemical structure is evident in figure 4.2, in which W_c versus F is compared for different C_6 hydrocarbons.

The dependence of W_c on F is smaller for monomers that tend to polymerize in plasma. The values of W_e/FM, as well as its dependence on F, vary with the chemical nature of the monomers. Thus, monomers with aromatic structure, $C{\equiv}C$ bonds, or $C{\equiv}N$ bonds have low values of W_e/FM. Monomers with cyclic structure and/or double bonds and Si compounds also have low W_e/FM values. Straight-chain hydrocarbons have much higher values of W_e/FM [32].

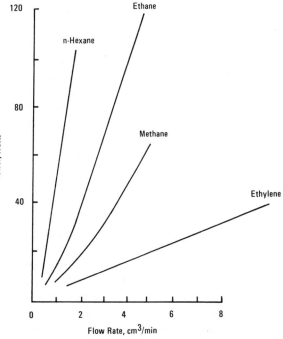

Figure 4.1. Power versus flow rate for hydrocarbons. *Source:* Adapted from H. Yasuda and T. Hirotsu, *J. Polym. Sci., Polym. Chem. Ed., 16,* 743 (1978). Copyright © 1978, John Wiley & Sons, Inc. Reprinted by permission of John Wiley & Sons, Inc.

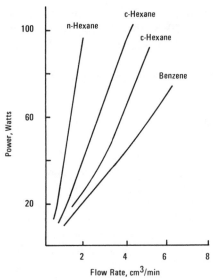

Figure 4.2. Power versus flow rate for hydrocarbons. *Source:* Adapted from H. Yasuda and T. Hirotsu, *J. Polym. Sci., Polym. Chem. Ed., 16,* 743 (1978). Copyright © 1978, John Wiley & Sons, Inc. Reprinted by permission of John Wiley & Sons, Inc.

The reason why W_e/FM increases is related to the hydrogen yield of monomers, that is, the "reactive" nature of polymer-forming plasma. The higher the value of γ and of F, the higher is the total number of molecules in the plasma gas phase, thus requiring more energy to sustain a glow discharge.

It is noted that hydrogen requires a relatively high energy to maintain glow discharge. The dependence of effective discharge power on flow rate for H_2, N_2, and Ar is illustrated in figure 4.3 [32]. The values of W_e/FM for gases and organic compounds are listed in table 4.4 [32].

Tetrafluoroethylene belongs to the group of monomers that is very sensitive to discharge power, and significant decomposition of this compound at high wattage drastically decreases deposition rates [33,34]. The initial increase in deposition rate with increasing discharge power is related to the development of full glow. Subsequent monomer decomposition causes a decrease in deposition rate after it has reached a maximum. These phenomena have been explained in terms of W/FM.

Tetrafluoroethylene requires a small wattage. W_e/FM is approximately 3.3 $\times\ 10^{-2}$ kWhr/g. When the value of W/FM exceeds 13×10^{-2} kWhr/g, decomposition of the monomer becomes evident. Therefore, at low flow rates that result in a high value of W/FM, no polymer deposition is observed except at very low wattage. This example may be typical for polymer deposition, in that it occurs only within a limited range of W/FM. For the case of tetrafluoroethylene, this effective range lies between 3.3×10^{-2} and 13×10^{-2} kWhr/g. The upper limit for hydrocarbons may be beyond the practical range

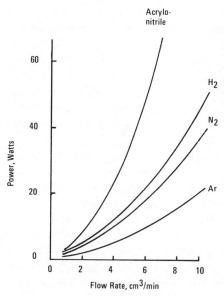

Figure 4.3. Power versus flow rate for acrylonitrile, H_2, N_2 and Ar. *Source:* Adapted from H. Yasuda and T. Hirotsu, *J. Polym. Sci., Polym. Chem. Ed., 16,* 743 (1978). Copyright © 1978, John Wiley & Sons, Inc. Reprinted by permission of John Wiley & Sons, Inc.

of plasma conditions. For low values of W_e/FM the deposition rate evidently increases with wattage. The significance appears to be in the need for defining the type of plasma polymerization by W/FM values.

F. Inductive versus Resistive Plasma

A concern that has been expressed in the literature pertaining to the interpretation of the mechanisms involved in plasma polymerization is whether

Table 4.4. Values of W_e/FM for gases and monomers

Gas or monomer	W_e/FM
H_2	56.8
N_2	3.7
Ar	1.5
Group 1 monomers	5.0
Group 2 monomers	5.0
Group 3 monomers	18.0

Note: Group 1 monomers = aromatics, C≡C, and C≡N bonds. Group 2 monomers = cyclic structures, double bonds. Group 3 monomers = straight-chain hydrocarbons.
Source: H. Yasuda and T. Hirotsu, *J. Polym. Sci., Polym. Chem. Ed., 16,* 743 (1978). Copyright © 1978, John Wiley & Sons, Inc. Reprinted by permission of John Wiley & Sons, Inc.

plasma polymerization carried out in the tail-flame part of inductively coupled r.f. discharge is the same as the polymer deposition onto the electrode surface in a resistively coupled discharge of the same r.f. power source. As an example, in the inductively coupled plasma at higher voltage, the polymer deposition is linearly related to monomer feed and is independent of wattage [35]; if both are within reasonable flow rates in a resistive discharge, polymer deposition rate depends on the monomer flow rate in a different manner, in that the deposition rate decreases with flow rate after a maximum has been reached [36,37,38,33].

In the inductively coupled plasma under full glow conditions, the rate of deposition is a function of monomer molecular weight [35], in that monomers with higher molecular weight exhibit a larger deposition rate. This tendency does not appear so evident under resistively coupled discharge [38]. Nevertheless, there seems to be no convincing evidence that the nature of the polymerization is significantly different in these two systems; that is, it may be the same phenomenon in a slightly different condition.

In plasma polymerization of acrylonitrile, a thorough investigation disclosed that the chemistry involved in plasma polymerization is largely determined by the character of the glow [39] and that comparison of two monomers should be conducted at comparable "glow character" instead of comparable flow pressure or discharge power. This means that when the rate of flow is increased at constant wattage, the value of the composite power parameter W/FM (see previous section) is decreasing. When this value decreases beyond a threshold value, the glow character and thus the polymer deposition rate change and begin to deviate from that of full glow conditions in an electrodeless discharge. On the other hand, when the wattage was increased as the flow rate increased, the situation was found to be the same as that observed in the electrodeless discharge, thus indicating that parameters such as varying flow rate at constant wattage or varying wattage at constant flow rate do not sufficiently describe conditions under which results should be compared.

It appears that comparisons between inductively coupled and resistively coupled systems are more realistic and result in similar deposition mechanisms if they are conducted under full glow or under similar glow condition.

3. Plasma Polymerization of Hydrocarbons

A. Saturated Aliphatic Hydrocarbons

The formation of organic polymers from hydrocarbons under the influence of electrical discharges had already been reported over a century ago by de Wilde [40], who obtained films from acetylene for which no solvent could be found. More recently, the composition of the polymers derived from excitation of hydrocarbon vapors were found to be as varied as their modes of formation. The decomposition of methane in a glow discharge operated at liquid air temperatures produced a conventional polymethylene, however [7]. On the other hand, it has been shown that both at 30 MHz [41] and 2450 MHz discharge [42] and ambient temperatures, low-molecular-weight saturated paraffins produced polymers having a C:H ratio of 2:3.

Conventional polymerizations of hydrocarbons normally require the presence in the monomer of some functional groups such as double bonds. Under plasma conditions, however, virtually any organic or organometallic vaporizable species can be polymerized. Here, the reaction is initiated by collisions between energetic free electrons in the glow discharge with the monomer molecules to generate active species. It has been shown [19,43,44,38] that these active species are predominantly free radicals and, to a lesser extent, ions and excited molecules. These then react either with themselves or with neutral molecules to form polymers both in the gas phase and on the nearest surfaces in the reactor.

Low background of air in the reaction system may result in polymers containing nitrogen and oxygen in their network structure. These have indeed been detected by emission spectrographic techniques as CN and CO, respectively [45]. Thus, free radicals and ionic species have been identified by a mass spectrometer sampling station [46,47,48]: in addition, methane, CH_4, the classic organic monomer, is also known to form under these plasma conditions. The dominant ions, identified from a capacitively coupled r.f. discharge [34], were C^+, CH^+, CH_2^+, $C_2H_2^+$, and $C_2H_3^+$. The system contained virtually no ions composed of more than three carbon atoms. Ions containing two or three C atoms were shown to result from ion-molecule interactions rather than from primary ionization of C_2 neutrals. The rate of polymer deposition was found to be greatly dependent upon both pressure and the location in the reactor. A schematic representation of the resistively coupled r.f. discharge used in these experiments (figure 4.4) [48] distinguishes between three regions of chemical interest: the dark space, R, between luminous plasma bulk and the r.f. electrodes; the container wall, W;

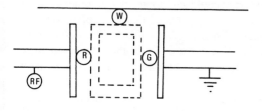

Figure 4.4. Resistively coupled r.f. discharge. *Source:* Adapted from M. J. Vasile and G. Smolinsky, *Int. J. Mass Spectrom. Ion Phys.*, *16*, 137 (1975).

Table 4.5. Percentage of C_1 ions

	Deposition rate for three regions (in Å/min)		
Pressure (torr)	R	G	W
0.8	11	11	3
0.5	16	16	9
0.3	31	26	11

Source: M. J. Vasile and G. Smolinsky, *Int. J. Mass Spectr. Ion Phys.,* *16*, 137 (1975).

and the ground electrode, G. Table 4.5 illustrates the rates of deposition for these three regions. It can be seen that the rates decrease as the pressure increases and that the rate of deposition on the walls is considerably lower than that on the electrodes.

These same authors [47] also showed that the CH_3^+ ions were by far the dominant species among the C_1^+ ions listed above. Among the identified C_2^+ ions, $C_2H_2^+$, $C_2H_3^+$, $C_2H_4^+$ and $C_2H_5^+$, the $C_2H_3^+$ species was found to prevail. Polymerization was found to occur chiefly on surfaces rather than in the gas phase, indicating that the chemical species arriving at the surface are the major determinants in the polymerization process.

The neutral species in the plasma were identified [49] as consisting essentially of molecular hydrogen and methane, along with small amounts of ethane, ethylene, and acetylene. Evidently, hydrogen can be dismissed as a progenitor of polymers in these experiments. Molecular methane, which impinges on the surface, is likely to encounter only radical species to form subsequently higher homologs. The two remaining neutral molecules, ethylene and acetylene, are assumed to function as polymer building blocks. The low concentrations of the two unsaturated monomers do not correlate with the measured high deposition rates, however. It appears, therefore, that the neutral species do not determine the rate of methane polymerization under these plasma conditions. It has indeed been shown that the C_1 and C_2 ions play a major role in this polymerization process.

A major source of free radicals is believed to result from neutralization at surfaces of such ions as $C_3H_3^+$, $C_2H_3^+$, CH_3^+, CH_2^+, and CH^+. The resulting radicals C_3H_3, C_2H_3, CH_2 and $CH\cdot$ should rapidly react with other radicals and unsaturated species. Therefore, it is postulated that this polymerization process proceeds via a free-radical mechanism. Since these radicals stem from the neutralization of the corresponding positive ions, the rate of polymer formation should then finally be determined by the rate of arrival of these ions.

Recent investigations [50] disclosed that in r.f. plasma the deposition rates of polymers from ethane, propane, and butane decreased in the order $C_2H_6 >$

$C_3H_8 > C_4H_{10}$. For these monomers the deposition rates also decreased with increasing flow rate. This general trend is consistent with that observed in the plasma polymerization of other monomers [38,51,52]. The interpretation is that the residence times of the reactive species were decreased by the increased flow rate and that some of the reactive species were swept away before polymerizations could take place. Methane differed from this trend in that its deposition rate first decreased to a minimum and then increased with flow rate.

Bell and his coworkers [49] also demonstrated that the deposition rates of these monomers were enhanced by increasing power input, which they explained by the generation of larger numbers of active species. Methane again proved to be the exception in that it decreased in deposition rate at high levels of power input. Plasma polymerization of ethane is slower than that of ethylene. It is assumed that an extra step is needed to effect the initial decomposition of ethane by electron impact to form ethylene first [38]

$$e + C_2H_6 = C_2H_4 + H_2 + e \qquad \text{(Eq. 4.26)}$$

followed eventually by acetylene formation

$$e + C_2H_4 = C_2H_2 + 2H \text{ (or } H_2) + e \qquad \text{(Eq. 4.27)}$$

and polymerization. The authors suggest that a similar mechanism may be responsible for propane and n-butane. For methane, however, initial decomposition to diradicals would be necessary to form ethylene from methane. Bell and his coworkers developed the "characteristic map," shown in figure 4.5, for the plasma polymerization of ethane. They had demonstrated earlier [13] that the form of plasma polymer from ethylene may be oil, film, or powder

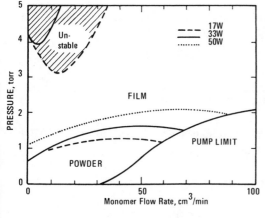

Figure 4.5. Characteristic map for plasma polymerization of ethane. *Source:* Adapted from H. Kobayashi, A. T. Bell, and M. Shen, *Macromolecules, 7,* 277 (1974); H. Hiratsuka, G. A. Kovali, M. Shen, and A. T. Bell, *J. Appl. Polym. Sci., 22,* 917 (1978). Reprinted with permission from *Macromolecules.* Copyright 1974, American Chemical Society.

and suggest that two competing processes may be operative; one is the rate of propagation in the gas phase, and the other is the rate of diffusion of active species from the plasma to the electrode surface. If the rate of polymerization is high, then the homogeneous reactions are prevalent in the gas phase to form powdery polymers. But if the diffusion rate is greater, then heterogeneous reactions on the surface will result in films or oils.

Acetylene is known to polymerize so rapidly that only powder can be formed. For ethylene all three forms of plasma polymer are possible. Since ethane polymerizes slowly in a glow discharge, films will be obtained in most cases. Increasing power input enhances the regions of powder formation because of the increased concentration of active species.

Following their outstanding experiments, Bell and his coworkers [50] developed the following empirical equation for the plasma polymerization of saturated hydrocarbons:

$$R_p = k_1 P^2 p \exp[-(k_2 p + k_3 F^{1/2})]$$ (Eq. 4.28)

where R_p is the deposition rate, P is the power input, p is the pressure, and F is the monomer flow rate; the k values are empirical constants.

B. Aromatic Hydrocarbons

Polymerization mechanisms of aromatic hydrocarbons, such as benzene, toluene, and so forth have been investigated in both high-energy r.f. discharges and high-energy microwave discharges. Under high-energy r.f. discharges of benzene, the main reactive species found are H, C_2H and, to a lesser extent, C_6H_5 and C_6H_4 radicals, that is:

$$C_6H_5 \rightarrow 3\ C_2H + 3H$$ (23)

while under high-energy microwave discharges the main emission consists in CH radicals and, to a lesser extent, C_2 and C_2H radicals:

$$C_6H_5 \rightarrow 6\ CH$$ (24)

$$C_6H_5 \rightarrow 3C_2H + 3/2\ H_2$$ (25)

Subsequent reactions of CH and C_2H radicals are known to yield initially olefinic moieties of the general form:

$$n\text{CH} \rightarrow -[\text{CH}{=}\text{CH}]_n-$$ (26)

$$n\text{C}_2\text{H} \rightarrow -[\text{C}{=}\text{CH}]_n-$$ (27)

Polymer crosslinking would then be brought about by a combination of free valences on neighboring chains.

In low-energy discharges, on the other hand, ring fragmentation is replaced by ring excitation [53,54], especially in r.f. discharges where the spectrum simply consists of the 2600 Å system of benzene. The observed formation of polymers with monosubstituted aromatic ring structure cannot be explained by condensation reactions of these excited molecules, however. A possible mechanism [55] could involve the immediate formation of acetylene, which is generated from benzene in electrical discharges [56,57], followed by its reaction with excited benzene molecules to form styrene:

$$C_6H_6{}^* + C_2H_2 \rightarrow C_6H_5 - CH{=}CH_2$$
$$\rightarrow -[CH{-}CH_2]_{\overline{n}}{-} \qquad (28)$$
$$\underset{\displaystyle C_6H_5}{\vert}$$

A similar mechanism involving the reaction with acetylene should apply to excited aromatic structures other than benzene.

In discharges of intermediate energy, there is a continuous variation, as a function of W/P ratio, of the amount of CH and H radicals formed in the discharges. It is generally observed that the reaction mechanism that applies to discharges of intermediate energy is a combination of those applicable to high-energy and low-energy discharges. Data indicate [53] that the results with toluene, ethylbenzene, and styrene suggest that the reaction mechanisms that apply to those monomers do not differ greatly from those applied to benzene.

Films obtained with values of W/P ratios larger than 200 were found to be insoluble [53] in most solvents, indicating a high degree of crosslinking. This is not unexpected, since the plasma is evidently more energetic the greater the value of W/P; 100 watts/0.2 torr = W/P = 500 would be considered a high ratio. The compounds found in the soluble films cover a broad range of molecular weights including diphenyl, terphenyl, quinquephenyl, and polymer films.

It has been reported [53] that the aromatic character of plasma-produced polymer decreases as the discharge becomes more energetic and that it is lower in microwave than in r.f. discharges. The aromatic character of polymers obtained in low-energy r.f. discharges of benzene is close to that of conventional polystyrene. Thus, polymers with large differences of aromatic character may be obtained in both microwave and r.f. discharges by simply varying the W/P ratios [53].

There is evidence [57,58] from IR spectroscopy that when the main aromatic bands are present, strong bands at 690 and 750 cm^{-1} show also the presence of pendant groups from an aliphatic or alicyclic chain segment.

Weaker bands at 690, 1020, and 1071 cm^{-1} had been assigned to a polybenzyl cyclopentene structure but could indicate a polystyrene structure as well [57].

The aforementioned mechanisms reintroduce some questions of polymerization in the gas phase versus initiation of polymerization on a reactor surface. Initiated chains formed in the gas-phase discharge cannot stay in the gas phase and must, therefore, deposit on the surfaces. Monomer molecules are adsorbed on these surfaces depending on pressure. If the pressure is relatively high, say 1 torr, there is enough monomer present to dissolve the migrating chain [59], and the degree of polymerization is found to be limited.

At low pressures, the surfaces are not covered with even a monomolecular layer of monomer, and the chain is found to continue growing. This trend is illustrated in figure 4.6, in which spectra obtained by gel permeation chromatography (GPC) show the effect of pressure on the formation of both polymer and low-molecular-weight aromatic compounds together with modal compounds. It is evident from this diagram that at higher pressures (within this pressure range) no polymer is formed, while the concentration of low-molecular-weight compounds increases.

Both cases are reminiscent of a mechanism related to emulsion polymerization, which would take place in the incompletely adsorbed layer of monomers in the electrode, with vacuum as the nonsolvent phase. Both conventional gas-phase as well as emulsion polymerization generally result in the formation of small spherical polymer particles, both being similar in size. Both

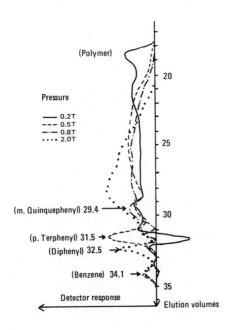

Figure 4.6. GPC spectra of soluble compounds deposited on electrodes at constant power (W/W$_0$ = 1.5) as function of pressure. *Source:* Adapted from M. Duval and A. Théorêt, *J. Appl. Polym. Sci., 17,* 527 (1973). Copyright © 1973, John Wiley & Sons, Inc. Reprinted by permission of John Wiley & Sons, Inc.

polymerization mechanisms have similar reaction rates. Ion-molecule reactions in gas phase, as well as the rate of arrival of monomer molecules to a surface, are known to increase with pressure.

Polymerization of benzene in both microwave and radio-frequency discharges [60] has shown that polymer structures obtained, as well as the composition of the plasma, are markedly influenced by the plasma energy level and the discharge frequency. The latter is particularly pronounced at high energy levels.

C. Unsaturated Hydrocarbons

Experimental data outlined in the previous section suggest that plasma polymerization of ethane is initiated by the formation of ethylene followed largely by decomposition to acetylene (reactions 26 and 27). In analogy with the photosynthesis of ethylene and the polymerization of ethylene by gamma radiation, the first stages of plasma polymerization of ethylene are also assumed to follow reactions 26 and 27. This assumption is supported by the observation that large amounts of acetylene are formed in the plasma [42,61].

The atomic hydrogen formed in the reaction

$$e + C_2H_4 \rightarrow C_2H_2 + 2H + e \tag{29}$$

may be converted into ethyl, methyl, and vinyl radicals [62,63].

$$H\cdot + C_2H_4 \rightarrow C_2H_5{}^* \rightarrow C_2H_5\cdot \tag{30}$$

$$H\cdot + C_2H_5 \rightarrow C_2H_6{}^* \rightarrow 2CH_3\cdot \tag{31}$$

$$H\cdot + C_2H_2 \rightleftharpoons C_2H_3{}.{}^* \rightarrow C_2H_3\cdot \tag{32}$$

Further reaction of these radicals could yield larger radicals and continue to grow in a gas-phase polymerization.

It has been suggested [1] that powder particles are formed in this manner by a vigorous polymerization of ethylene. At the low pressures generally employed, a fraction of the free radicals can be expected to diffuse in the gas phase to the surfaces that contain the plasma to induce reactions 5-11.

In addition to the radicals diffusing to the surface, a certain number may also be formed through the impingement of energetic ions and the adsorption of UV radiation produced by the plasma:

$$I^* + S—M \rightarrow S—R\cdot + H\cdot + I \tag{33}$$

$$h\nu + S—M \rightarrow S—R\cdot + H\cdot \tag{34}$$

where I* and I are ions in the activated and deactivated states. Subsequently, polymerization is expected to grow through reactions 5–11.

The H/C ratio of both the film and the powder lies between 1.3 and 1.5 [8,41,42]. This ratio may be explained by the polymerization of equivalent amounts of ethylene and acetylene. Alternatively, the polymer can first be formed with a H/C ratio of approximately 2, followed by hydrogen abstraction from the polymer by UV adsorption and collisions with energetic species.

It has been shown in the previous section that the plasma deposition rates of polymers from ethane, propane, and butane decreased in the order $C_2H_6 > C_3H_8 > C_4H_{10}$. This same order is also observed in the plasma polymerization of ethylene, propylene, and butylene. Acetylene polymerizes so quickly with the evolution of small amounts of hydrogen that the system pressure remains very low. In the presence of nitrogen, abnormally high deposition rates were measured [64]. This effect was attributed to a shifting of the deposition distribution away from the site of the probe. At low monomer flow rates, the distribution of polymer deposition is quite uniform. At higher flow rates, the site of maximum deposition tends to shift more toward the downstream side of the glow. Although the distribution of polymer deposition becomes less uniform, the ratio of the maximum value/average deposition rate does not change markedly with increasing monomer flow rate.

Addition of a carrier gas (at constant acetylene flow rate) alters the deposition patterns drastically [64], however. Since nitrogen and hydrogen are known to copolymerize with acetylene in a glow discharge, the gases should contribute to the enhancement of deposition rates. The contribution of hydrogen is probably negligible, but that of nitrogen is considered to be significant, since the molecular weight of N_2 [28] is close to that of acetylene [27]. The effect of carrier gas type on acetylene deposition rate is illustrated in table 4.6.

In plasma polymerization of ethylene the effect of carrier gas on the deposition rate is less pronounced (table 4.7). The gas flow rate was 2.5 cm³/min.

Table 4.6. Effect of carrier gas on acetylene deposition rate

Carrier gas	Molecular ratio gas/acetylene	System pressure μm Hg	Deposition rate, Å/min
—	—	8	~400
H_2	1.3	70	~840
Ar	0.8	8.8	~850
N_2	1.0	45	~1500

Source: H. Yasuda and T. Hirotsu, *J. Polym. Sci., Polym. Chem. Ed.,* *16,* 229 (1978). Copyright © 1978, John Wiley & Sons, Inc. Reprinted by permission of John Wiley & Sons, Inc.

Table 4.7. Effect of carrier gas on ethylene deposition rate

Carrier gas	Molecular ratio gas/ethylene	System pressure μm Hg	Deposition rate, Å/min
—	—	44	~280
H_2	1.1	87	~380
Ar	0.8	110	~420
N_2	0.9	96	~610

Source: H. Yasuda and T. Hirotsu, *J. Polym. Sci., Polym. Chem. Ed., 16,* 229 (1978). Copyright © 1978, John Wiley & Sons, Inc. Reprinted by permission of John Wiley & Sons, Inc.

In these experiments Yasuda and Hirotsu [64] introduced the gas mixture in the center portion of the reactor so that the monomer entered the r.f. glow discharge directly and therefore was able to react instantaneously. Such an arrangement may alter the modes by which monomers are converted into their active species and the reactivity of the monomer in the plasma state as well.

Reflecting the difference of reactivities, ethylene displays a more uniform distribution of deposition than does acetylene. Ethylene, which is less reactive than acetylene, travels randomly through the reaction vessel before settling down to form polymer deposits, while acetylene tends to deposit rapidly near the monomer inlet.

The diffusional displacement distance of a gas is given by Einstein as

$$X^2 = 2Dt \tag{Eq. 4.29}$$

where X is the one-dimensional displacement by a particle within time t in a gas of a diffusion coefficient D. At a low pressure, D can be calculated [65] from the diffusion coefficient D_0 at the standard state ($p_0 = 760$ mm Hg and $T_0 = 273°$K) by

$$D = D_0 \, (T/T_0)^n \, (p_0/p) \tag{Eq. 4.30}$$

where n lies between 1.75 and 2.

Thus, the diffusional displacement distance is shorter for higher-molecular-weight molecules of low D_0 value and for the system at higher pressure. Hence a higher reactivity of monomer plasma should lead to a narrower distribution of polymer deposition. Carrier gas was shown [64] to cause a narrowing of polymer distribution, this effect decreasing in the order $N_2 > Ar > H_2$.

It should be noted that some gases that do not polymerize alone in plasma, such as H_2, N_2, H_2O, and CO, are known to copolymerize with other monomers such as acetylene and ethylene [66,67]. Therefore, the sequence $N_2 > Ar > H_2$ may not be related solely to a mass effect.

Yasuda and his coworkers [64] point to the clear distinction between rate of polymerization and rate of polymer deposition. If there is no complete conversion of monomer to polymer, the distribution of polymer deposition cannot be related to an absolute polymerization rate. Other factors, such as the relationship between reactor volume and glow volume, V_R/V_G, equal 1 if both V_R/V_G and the flow pattern are defined.

Flow rate is to be understood as the flow in terms of number of moles but is not directly related to the velocity of a gas flow. In a noncompressive system such as a liquid, the flow rate is directly proportional to the flow velocity. In a vacuum, the flow rate in the context of flow velocity is given by F/p, where F is the flow rate in cm^3 (STP) and p is the system pressure in atm. Thus, the same flow rate, that is, 10 cm^3 (STP)/min at 0.1 mm Hg, has a velocity 100 times higher at 10 mm Hg.

It should also be noted that the linear viscosity of flow is dependent upon the cross-sectional area of the system in which the flow takes place. Therefore, the linear velocity of a flow can be described by $(F/A)_p$, where A is the cross-sectional area in cm^2. For instance, the value of $(F/A)_p$ for a flow of 50 cm^3 (STP)/min entering a bell jar of 50 cm diameter at 1 mm Hg, is 19.4 cm/min, while the value of $(F/A)_p$ for a flow of 5 cm^3/min maintained in a tube of 5 cm diameter at 0.1 mm Hg is 19-36 cm/min. Evidently, 50 cm^3/min in the bell-jar system is a much slower flow than that of 5 cm^3/min in the tube system under these conditions, although F is ten times greater.

In describing the velocity of monomer feed into the reactor, the ratio F/V_g must be taken into consideration, while the residence time of a flow system relates to $(F/A)_p$. Therefore, in the absence of the system description, a value of F alone has little value beyond being a parameter for a given system. Yasuda and Hirotsu [64] point out that these factors are generally not considered in discussions of plasma polymerizations appearing in the literature.

In plasma polymerization, the mass transport is important, regardless of reactor design. In addition, the mode of energy input to maintain a given plasma state and its relative position to the location of polymer deposition is important in plasma polymerization. Yasuda and his coworkers [68] have recently investigated the deposition rate and deposition-rate distribution at the electrode as well as at a substrate placed midway between electrodes as a function of pressure and flow rate of ethylene, power, frequency, and the presence of magnetic enhancement. These investigators employed a magnetic field to confine the plasma in the inner electrode gap [69]. Conditions for best polymerizations were described as follows.

Low pressure–low flow rate is preferable to high pressure–high flow rate to provide improved monomer conversions at up to 70 W power input. Low pressure also yields higher total polymer deposition on a substrate midway between electrodes, as compared with the deposition on the electrodes themselves (at the experimental frequencies of 60 Hz and 10 KHz). Adhesion of polymer to aluminum improved when magnets are employed. Under the preferred low-pressure conditions, a.c. operation should be avoided due to excessive heat buildup of the electrodes.

It is generally recognized that electron-impact ionization is the major source for the generation of charged species in a glow discharge. Another important ionization mechanism is the direct ionization by collision with sufficiently energetic, metastable neutral species. This process, described previously as the Penning ionization, becomes a major mechanism whenever certain gases are employed as carriers in the polymerization. For He, Ne, and Ar the excitation energies of the lowest triplet states are 19.80, 16.62, and 11.55 eV, respectively. Although the lowest triplet state in "active nitrogen" does not exist because of the efficient quenching of it by nitrogen atoms [21], "active nitrogen" is unique because of its extensive vibrational excitations, which continue for prolonged periods beyond the discharge zone and therefore lead to considerable chemiionization [20]. The ionization potential for most neutral gases is approximately 15 eV. Thus, both the effect of carrier gases, as mentioned earlier and throughout this book, and the aforementioned parameters, should be considered in terms of Penning ionization.

In the polymerization of ethylene, the resulting polymer may be obtained in the form of oil, powder, or film [13]. At low pressures and low monomer flow rates, the polymer is present both as a powder and a film. At high pressures and high flow rates, an oily film is obtained. Only at low pressure and high flow rate is a solid, pinhole-free film deposited. If the reactor pressure is high and the flow rate is low, the discharge becomes unstable [8].

Depending on the experimental conditions, many monomers have been obtained in all three forms. The oily polymers are generally found to contain pendant groups, while both the films and powders are crosslinked. Their structure will be discussed in a separate section.

D. Vinyl Monomers

Styrene monomer has recently been selected for a detailed study of the polymerization mechanism of vinyl-type monomers in a low-temperature r.f. plasma [70]. Polymer deposition is shown to increase with power in table 4.8.

Table 4.8. Rate of polystyrene formation and monomer conversion at various power levels

Power (watts)	dP/dt (g/hr)	% Monomer conversion
8	0.019	30–35
13	0.035	60–65
18	0.047	—
22	0.053	80–85
36	0.057	100

Source: L. F. Thompson and K. G. Mayhan, *J. Appl. Polym. Sci., 16,* 2317 (1972). Copyright © 1972, John Wiley & Sons, Inc. Reprinted by permission of John Wiley & Sons, Inc.

The rate of polymer formation follows an Arrhenius-type plot (figure 4.7) of the type

$$\ln \text{(rate of polymer formation)} = Ae^{-E_a/K'W} \qquad \text{(Eq. 4.31)}$$

where $-E_a$ is the apparent activation energy, W is the power input into the system in watts, and K' is a constant that includes time so that the product $K'W$ can be compared to the product RT from a standard Arrhenius expression.

In the study of the initiation step, the investigators introduced nitrogen dioxide, an efficient free-radical scavenger, into the system along with the monomer. The reaction should be retarded or inhibited if the initiation occurs via a free-radical mechanism. Although NO_2 will be excited by the plasma into ionic and free-radical species, it was argued that an ion of NO_2 is short-lived at a pressure of 25 microns because of recombinations through wall collision. Therefore, some of the NO_2 should exist in its free state in the gas phase to serve as a scavenger species. The scavenger gas did not affect the rate of polymerization to a measurable extent. This would indicate that the initiation step does not occur via either a free or an ion radical.

In a second step the investigators made use of an electrostatic ion-deflection

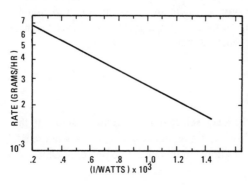

Figure 4.7. Arrhenius-type plot for styrene polymerization. *Source:* Adapted from L. F. Thompson and K. G. Mayhan, *J. Appl. Polym. Sci., 16,* 2317 (1972). Copyright © 1972, John Wiley & Sons, Inc. Reprinted by permission of John Wiley & Sons, Inc.

device to detect the existence of ionic species in the plasma. The greater part of the plasma was found to be deflected toward the negative electrode, indicating that cations represented the largest fraction of charged species. Experiments conducted at 22 watts for 12 hours showed that over 90% by weight of the polymer was deposited on the negative electrode and no polymer was formed downstream from the electrodes. These results indicated that the reaction is initiated through a cationic mechanism.

On the basis of these findings, the researchers discussed the initiation steps as though ions or ion radicals could cause initiation. (In the light of the scavenger gas experiments, however, they felt that the ion radical is less likely to be involved.) Thus, molecules are excited in an r.f. plasma by elastic collision with accelerated electrons or accelerated ions. The following series of events is contemplated:

$$M \xrightarrow{\Delta E} M* \tag{35}$$

$$M* + e \rightarrow M^+ + 2e \tag{36}$$

$$M* + e \rightarrow M^- \tag{37}$$

$$M* \rightarrow M^{\overset{+}{\cdot}} + e \tag{38}$$

$$M* + e \rightarrow M^{\overline{\cdot}} \tag{39}$$

$$M* \rightarrow p^+ + Q^- \tag{40}$$

where M = monomer molecule; ΔE = the energy from plasma (an accelerated particle, usually an electron); M* = excited monomer; M^- = anionic monomer molecule; M^+ = cationic monomer molecule; $M^{\overset{+}{\cdot}}$ = cation radical; $M^{\overline{\cdot}}$ = anion radical; e = electron; p^+, Q^- = styrene fragments. Step 40 may be fragmented as shown below.

$$\tag{41}$$

The second series of initiation steps is possible if the active monomer molecule, capable of propagation, is formed upon interaction with an accelerated particle represented as in electrons.

$$M + e \xrightarrow{\Delta E} M^+ + 2e \tag{42}$$

$$M + e \xrightarrow{\Delta E} M^- \tag{43}$$

$$M + e \xrightarrow{\Delta E} M^{\pm} + 2e \tag{44}$$

$$M + e \xrightarrow{\Delta E} M^{\mp} \tag{45}$$

The propagation steps were outlined to produce long-chain polymer:

$$M^+ + M \rightarrow MM^+ \tag{46}$$

$$MM^+ + nM \rightarrow M(M)_a M^+ \tag{47}$$

Crosslinking and branching were also suggested to occur via a cationic mechanism:

$$MMM(M)_n M \xrightarrow{\quad \Delta E \quad} MMM(M)_n M^+ \tag{48}$$

$$\begin{array}{c} MMM(M)_n M^+ + M \longrightarrow \begin{array}{c} MMM(M)_n M^+ \\ | \\ M_+ \end{array} \end{array} \tag{49}$$

$$\begin{array}{c} \begin{array}{c} MMM(M)_n M^+ \\ | \\ M_+ \end{array} + nM \longrightarrow \begin{array}{c} MMM(M)_n M^+ \\ | \\ M \\ | \\ M \\ | \\ (M)_n \\ | \\ M_+ \end{array} \end{array} \tag{50}$$

From these and other experiments, the investigators drew the following conclusions concerning the polymerization of styrene and other vinyl-type monomers.

These monomers can be converted to polymers in an inductively coupled r.f. plasma at any site in the reactor, depending on experimental parameters such as power level, monomer input rate, and time.

A kinetic treatment of the rate fits an Arrhenius expression.

1–2% of the polymer was soluble and consisted of only oligomers.

The plasma polymer was thermally much more stable than the linear polystyrene.

The plasma polymer was highly crosslinked.

The material did not possess any long-range order that could contribute to its insolubility in organic solvents.

Scanning electron micrographs of the films showed clearly that the polymer formation is not solely an adsorption-controlled process. Polymer is also formed in the vapor phase.

The polymerization proceeds through a cationic mechanism.

Crosslinking of the polymer occurs during the polymer formation.

The view of Thompson and Mayhan in 1972 that styrene polymerization in plasma proceeds essentially via a cationic process was supported by Westwood [71] in 1971 in low-temperature r.f. polymerization of other vinyl monomers, such as vinyl chloride and vinyl fluoride, methyl methacrylate and vinylidene chloride.

Westwood suggests from the experimental evidence that glow-discharge polymers from these monomers are formed on the electrode surface by interaction of positive gas-phase ions with adsorbed monomer molecules. He, too, cites as evidence that a monomer vapor is deposited in a direct-current system to form a polymer exclusively on the cathode. This implies that positive ions play a vital part in the production of polymer. He suggests that the electrons, in being accelerated across the anode fall, are inefficient in initiating polymerization on the anode. He supports this view by grid experiments. With zero d.c. potential between the grid and the counter electrode, a certain amount of polymer collected on the counter electrode. Application of a negative potential to the counter electrode, which presumably accelerates positive ions out of the plasma, increases the rate of polymer formation on the electrode. On the other hand, only a small positive potential is required to reduce the rate of polymer formation to zero. In the case of r.f. discharge, equal numbers of positive ions and electrons will diffuse out of the plasma, the rate of arrival of positive ions at the electrode surface was shown to be of the same order of magnitude as the rate of incorporation of molecules into the growing polymer.

Westwood's data indicate that for vinyl chloride, with an electrode temperature of 15°C, a current density of 2 mA·cm^{-2} and at 1 torr pressure, the polymer deposition rate is 90 $\mu g \cdot cm^{-2}$ min^{-1}, resulting in a rate of incorporation of 1×10^{16} molecules cm^{-2} sec^{-1}. Under these conditions the number of gas molecules colliding with the surface is 3.8×10^{20} molecules cm^{-2} sec^{-1}. Calculation indicates that the number of positive ions incident on the surface is $\sim 10^{16}$ cm^{-2} sec^{-1}. Assuming that all these ions adhere to the surface, then their rate of arrival is of the same order of magnitude as the rate of incorporation of molecules into the polymer structure. Therefore, in view of such a low surface coverage, each ion reacts with only a small number of adsorbed molecules. These then would be available for continued initiation processes by impingement with molecules subsequently arriving at the surface.

Interestingly, Westwood, in support of the cationic polymerization mechanism, cites a comparison with radiochemistry, the same example that, Bell and his coworkers argue, supports the free-radical mechanism [8, equations 3 and 4].

Westwood calculated the efficiency of the glow-discharge polymerization process in terms of G values. The G_x value of a process is the number of "x" events occurring per 100 eV of energy absorbed. For methyl methacrylate at $-20°$ and a pressure of 1.9 torr, for which the p/p_0 value is 0.9, the polymer yield was 0.0071 g for a power input of 5.51 W. This means that 2.4×10^{17} molecules/sec of monomer were incorporated into the polymer for an energy absorption of 3.4×10^{19} eV; therefore, the G value for monomer to polymer was 0.7.

This value is considerably lower than the G values found in radiation-induced polymerizations in liquids. Thus, even in the relatively favorable case of a monomer close to its point of condensation on the electrode surface, the polymerization has a low G value. Westwood suggests as an explanation that the rate of energy input in a glow discharge is 10^6 times greater than that used in radiation-induced polymerization reactions. Chain termination rather than propagation will be favored by these high-energy input rates. Since an equal number of positive ions and electrons will diffuse out of the r.f. plasma, the termination step probably involves recombination of a polymer ion, or radical, on the electrode surface, with an electron incident from the gas phase. In table 4.9 the rates of formation of various vinyl monomers are compared with those of ethylene. Acetylene lies in polymer deposition rates between vinyl chloride and vinyl fluoride [72].

The observation that hydrocarbon halides in general exhibit a higher rate of polymer deposition when compared with unhalogenated simple hydrocarbons was rationalized by Bell and his coworkers [72] by considering the ease with which free radical and unsaturated species can be formed in the plasma. IR spectra indicate that the structures of plasma-polymerized vinyl chloride and vinyl fluoride are in many respects similar to the plasma-polymerized hydrocarbon.

Table 4.9. Rates of formation of glow-discharge polymers

| | Discharge conditions | | | |
	Current (mA/cm^2)	Pressure (torr)	Temp. (°C)	Rate (Å/sec)
Monomer				
Vinyl acetate	2.43	0.99	0	40
Methyl methacrylate	2.43	1.96	0	130
Vinyl chloride	1.53	0.99	0	160
Vinylidene fluoride	3.73	0.82	20	260
Vinyl fluoride	2.01	0.63	0	85
Styrene	1.28	0.98	20	40
Tetrafluoroethylene	2.42	0.95	20	14
Ethylene	2.01	0.63	20	4
Vinylidene chloride	2.43	0.63	20	86

Source: A. R. Westwood, *Europ. Polym. J.*, *7*, 363 (1971). Reprinted with permission from *European Polymer Journal*. Copyright 1971, Pergamon Press, Ltd.

It is noted, however, that the IR-spectrum of plasma-polymerized tetrafluoroethylene does not resemble that of conventional polytetrafluoroethylene (PTFE). Both the evidence from photolysis studies and consideration of the heats of reaction [72] suggest that the primary active species in the initiation process is the CF_2, a free radical derived directly from TFE. This is in direct contrast with ethylene, where the primary active species is believed to be $H \cdot$. The heats of reaction associated with both processes have been reported by Bell and coworkers [72] (table 4.10). It is immediately apparent from the heats of reactions that difluoromethylene is the preferred dissociation product for TFE and that for ethylene, the preferred dissociation products are acetylene and molecular hydrogen. The generally observed higher rate of deposition of halogenated hydrocarbons as compared with the unhalogenated hydrocarbons appears to be related to the presence of halogen atoms in the gas phase. This is dramatically demonstrated in figure 4.8 by the effect of Freon 12 on the deposition rates of ethane and ethylene [72]. Similar results were

Table 4.10. Heats of reaction for fluorocarbons and hydrocarbons

Plasma components	Dissociation products	Heat of reaction (Kcal/mole)	
$e + C_2F_4 \longrightarrow$	$C_2F_2 + F_2 + e$	$\Delta H = 162.4$	(51)
$\longrightarrow$	$C_2F_2 + 2F \cdot + e$	$\Delta H = 200.2$	(52)
$\longrightarrow$	$2CF_2: + e$	$\Delta H = 70.4$	(53)
$e + C_2H_4 \longrightarrow$	$C_2H_2 + H_2 + e$	$\Delta H = 41.7$	(54)
$\longrightarrow$	$C_2H_2 + 2H \cdot + e$	$\Delta H = 145.0$	(55)
$\longrightarrow$	$2CH_2: + e$	$\Delta H = 171.7$	(56)

Source: H. Kobayashi, M. Shen, and A. T. Bell, *J. Macromol. Sci., Chem., A8*(8), 1345 (1974). Reprinted by courtesy of Marcel Dekker, Inc., N.Y.

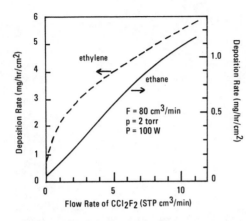

Figure 4.8. Effect of CCl_2F_2 addition on the deposition rates of hydrocarbons. *Source:* Adapted from H. Kobayashi, M. Shen, and A. T. Bell, *J. Macromol. Sci., Chem.,* A8(8), 1345 (1974). Reprinted by courtesy of Marcel Dekker, Inc., N.Y.

obtained from addition of halogenated hydrocarbons to methane in a plasma polymerization. As an example, addition of 10 cm^3/min of Freon 12 to a flow of 50 cm^3/min of methane increases the polymer deposition rate fiftyfold over that for methane alone [72]. The resulting polymer was found to contain only very small amounts of chlorine. It is shown by the heat of reactions that methane requires considerably more energy to dissociate into free radicals than do halogenated hydrocarbons such as CF_2Cl_2 and CH_3Cl (table 4.11).

The expectation that by this initiation process, the heats of reaction of CF_4 should have little effect on methane polymerization, was experimentally confirmed [72]. In addition to forming free radicals by direct dissociation, the release of chlorine atoms (reactions 60 and 62) should also contribute to H abstraction from hydrocarbon intermediates and the growing polymer.

The mechanism of plasma polymerization of the monomer 2-vinylpyridine, (2-VP), which is known to undergo free-radical as well as anionic-catalyzed

Table 4.11. Heats of reaction for halogenated and nonhalogenated hydrocarbons

Plasma components	Dissociation products	Heat of reaction, ΔH (Kcal/mole)	
$e + CH_4$	$\longrightarrow CH_2: + H_2 + e$	105.9	(57)
	$\longrightarrow CH_3\cdot + H + e$	104.0	(58)
$e + CF_2Cl_2$	$\longrightarrow CF_2: + Cl_2$	74.0	(59)
	$\longrightarrow CF_2Cl\cdot + Cl$	78.5	(60)
$e + CH_3Cl$	$\longrightarrow CH_2: + HCl$	90.7	(61)
	$\longrightarrow CH_3\cdot + Cl\cdot$	84.4	(62)
$e + CF_4$	$\longrightarrow CF_2: + F_2$	179.5	(63)
	$\longrightarrow CF_3\cdot + F\cdot$	119.5	(64)

Source: H. Kobayashi, M. Shen, and A. T. Bell, *J. Macromol. Sci., Chem.,* A8(8), 1345 (1974). Reprinted by courtesy of Marcel Dekker, Inc., N.Y.

addition polymerization readily, was recently studied by Holovka [73]. He found that, in the 10 KH_z a.c. glow discharge, polymerization of 2-VP, polymer is found in significant amounts only on the electrode surfaces and very little on other surfaces near the glow region of the discharge. This observation suggests that polymerization is initiated and propagated only on the electrode surface rather than in the vapor phase.

The deposition rate at constant monomer pressure of 100 μm torr is clearly first order (figure 4.9). The overall rate expression for the formation of polymer on the electrode surfaces was given by

$$R = \frac{2270j(P_{2\text{-VP}} + 94)}{T}$$ (Eq. 4.32)

where R is in units of Å/min; the current density, j, in mA/cm^2; the pressure $P_{2\text{-VP}}$ in millitorr; and T in °C. This relationship was found to hold for $80 < P_{2\text{-VP}} < 180$ millitorr and was essentially unaffected by the presence of argon in discharge gases.

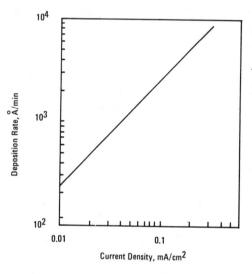

Figure 4.9. Rate of formation of PVP as a function of plasma current density. *Source:* Adapted from H. Kobayashi, M. Shen, and A. T. Bell, *J. Macromol. Sci., Chem. A8*(8), 1345 (1974). Reprinted by courtesy of Marcel Dekker, Inc., N.Y.

In a direct-current discharge, polymer was found only on the cathode. Therefore, Holovka suggests, ions are more effective than photons or electrons in initiating polymerization. This could be argued as evidence for cationic polymerization. The nitrogen in the pyridine ring would tend to form a complex with the ion and would thus reduce the ion's reactivity, however. A more likely possibility is suggested, namely, that the impinging ion strikes the surface and generates free-radical sites on the surface. These in turn initiate polymerization or recombine with other free-radical sites to form polymer crosslinks.

Fluorocarbon films, obtained by introducing perfluorobutene-2 into the afterglow of an r.f. argon plasma, were recently found to contain high concentrations of free radicals [74]. The spin concentration was estimated from ESR signals to be about 10^{20} spin·gram^{-1}. The signal decayed slowly, reaching half-intensity after 16 days at ambient temperature. The presence of free radicals is a frequently observed characteristic of films formed in plasma systems [75,76].

In a more recent investigation of r.f. plasma polymerization of 1,2- and 1,1-difluoroethylene, it was suggested [77] that a precursor to the polymer is based on fluoroacetylene, C_2F_2. Accordingly, interaction of a difluoroethylene molecule with electrons in the plasma will lead to the production of excited states of the neutral molecule, followed by ionizations to form radical cations in various electronic states.

The lowest-lying excited states of an ethylenic system originate in $\pi \rightarrow \pi^*$ excitations, and for electron-impact excitation, singlet and triplet states are available; the corresponding excitation energies are about 7.5 and 4.4 eV [78]. In an alternate hydrocarbon the perturbation of the system by replacement of a hydrogen by a fluorine substituent is small; therefore the transition energies to the singlet and triplet $\pi \rightarrow \pi^*$ excited states of difluoroethylenes are the same. The first ionization potential corresponds to removal of a π electron, and that of the isomeric difluoroethylenes is similar to that of ethylene (ethylene: 10.51 eV, 1,1-difluoroethylene: 10.48 eV).

The most likely initial products are thought to be the fluoroacetylene radical cation and hydrogen fluoride followed by the products shown below.

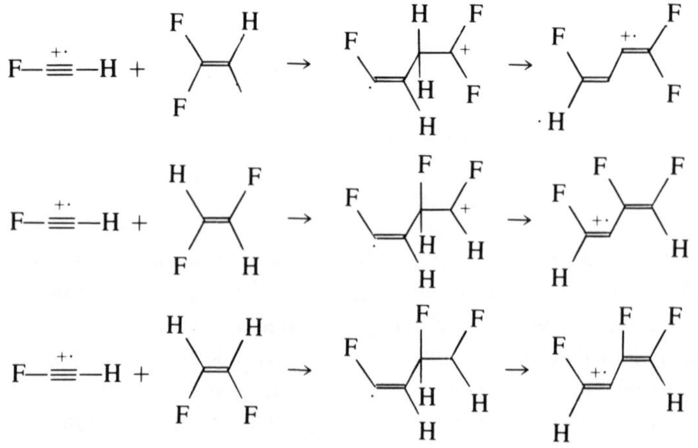

References

1. W. B. Kunkel. In *Plasma Physics in Theory and Application*, Ch. 10. McGraw-Hill: New York (1966).

2. J. A. Coffman and W. A. Browne. *Sci. Amer.*, 91 (June 1965).

3. J. D. Thornton. *Chem. Processing, 12*(2), S.6 (1966).

4. H. Suhr. In *Techniques and Applications of Plasma Chemistry*, ed. J. R. Hollahan and A. T. Bell. Wiley: New York (1974).

5. A. Charlesby. *Atomic Radiation and Polymers*. Pergamon Press: Oxford (1960).

6. P. M. Hay. In *Chemical Reactions in Electrical Discharges*, ed. B. D. Blaustein. Advances in Chemistry Ser., *No. 80*. Amer. Chem. Soc.: Washington, D.C. (1969).

7. L. M. Yeddanapalli. *J. Chem. Phys., 10*, 249 (1942).

8. H. Kobayashi, M. Shen, and A. T. Bell. Res. Contract, AD-762, 480, Office of Naval Res. (15 June 1973).

9. A. T. Bell. I & EC Fund., *11*, 209 (1972).

10. A. T. Bell and K. Kwong. I & EC Fund., *12*, 90 (1973).

11. P. Borell, A. Corvenka, and J. W. Turner. *J. Chem. Soc.* (B), 2293 (1971).

12. H. Mitsu, S. Machi, M. Hagiwara, and T. Kagiya. *J. Polym. Sci., A-1, 5*, 1073 (1967).

13. H. Kobayashi, A. T. Bell, and M. Shen. *J. Appl. Polym. Sci., 18*, 885 (1973).

14. S. C. Brown. *Introduction to Electrical Discharges in Gases*. Wiley: New York (1966).

15. A. R. Denary, P. A. Owens, and A. Crawshaw. *European Polymer J., 4*, 93 (1968).

16. C. H. Bamford, A. D. Jenkins, and J. C. Ward. *Nature*, London, *186*, 712 (1960).

17. T. Williams and M. W. Hayes. *Nature*, London, *209*, 769 (1966).

18. A. R. Denary, P. A. Owens, and A. Crawshaw. *European Polymer J., 4*, 93 (1968).

19. H. Yasuda and C. E. Lamaze. *J. Appl. Polym. Sci., 15*, 2277 (1971).

20. F. Kaufman. 153rd National Meeting of the Amer. Chem. Soc., Div. Fuel Chemistry, Miami Beach, Fla., April 1967, Preprints 11 (No. 1), 11. Amer. Chem. Soc.: Washington, D.C. (1967).

21. R. A. Young, *Can. J. Chem., 44*, 1171 (1966).

22. R. Liepins and K. Sakaoku. *J. Appl. Polym. Sci., 16*, 2633 (1972).

23. Y. Osada, A. T. Bell, and M. Shen. *J. Polym. Sci., Polym. Letters Ed., 16*, 309 (1978).

24. A. Chapiro. *Radiation Chemistry of Polymer Systems*. Interscience: New York (1962).

25. W. H. Seitzer and A. V. Tobolsky. *J. Amer. Chem. Soc., 75*, 755 (1953).

26. T. G. Majury. *J. Polym. Sci., 15*, 297 (1955).

27. F. A. Bovey. *Polymer Conformations and Configurations*. Academic Press: New York (1969).

28. A. T. Bell. In *Techniques and Applications of Plasma Chemistry*, ed. J. R. Hollahan and A. T. Bell. Wiley: New York (1974).

29. W. P. Allis. In *Handbuch der Physik*, Vol. 21. Springer Verlag: Berlin, p. 383 (1956).

30. D. K. Lam. In *Plasma Chemistry of Polymers*. Dekker: New York (1976).

31. B. D. Washo. In *Plasma Chemistry of Polymers*. Dekker: New York (1976).
32. H. Yasuda and T. Hirotsu. *J. Polym. Sci., Polym. Chem. Ed., 16*, 743 (1978).
33. H. Yasuda and T. S. Hsu. *J. Polym. Sci., Polym. Chem. Ed., 16*, 415 (1978).
34. H. U. Poll, M. Arzt, and K. H. Wickleder. *Eur. Polym. J., 12*, 505 (1976).
35. H. Yasuda and C. E. Lamaze. *J. Appl. Polym. Sci., 17*, 1519 (1973).
36. K. C. Brown. *Eur. Polym. J., 8*, 117 (1972).
37. K. C. Brown and M. I. Copsey. *Eur. Polym. J., 8*, 129 (1972).
38. H. Kobayashi, A. T. Bell, and M. Shen. *Macromolecules, 7*, 277 (1974).
39. H. Yasuda and T. Hirotsu. *J. Appl. Polym. Sci., 21*, 3139 (1977).
40. P. de Wilde. *Ber., 7*, 302 (1874).
41. F. K. McTaggert. *Plasma Chemistry in Electrical Discharges*. Elsevier: Amsterdam (1967).
42. F. J. Vastola and J. P. Wightman. *J. Appl. Chem., 14*, 69 (1964).
43. M. Duval and A. Théorêt. *J. Appl. Polym. Sci., 17*, 527 (1973).
44. H. Charchano. *J. Chem. Phys., 61*, 3634 (1974).
45. C. T. Wendel and M. H. Wiley. *J. Polym. Sci., A-1, 10*, 1069 (1972).
46. M. J. Vasile and G. Smolinsky. *Int. J. Mass Spectr. Ion Phys., 12*, 133 (1973).
47. M. J. Vasile and G. Smolinsky. *Int. J. Mass Spectr. Ion Phys., 12*, 147 (1973).
48. M. J. Vasile and G. Smolinsky. *Int. J. Mass Spectr. Ion Phys., 16*, 137 (1975).
49. G. Smolinsky and M. J. Vasile. In *Plasma Chemistry of Polymers*. Dekker: New York (1976).
50. H. Hiratsuka, G. Akovali, M. Shen, and A. T. Bell. *J. Appl. Polym. Sci., 22*, 917 (1978).
51. T. Reis, H. Hiratsuka, A. T. Bell, and M. Shen. *Natl. Bur. Stand. (U.S.) Spec. Publ. No. 462*, Washington, D.C., p. 230 (1976).
52. H. Kobayashi, A. T. Bell, and M. Shen. *J. Macromol. Sci., Chem., A10*, 123 (1976).
53. M. Duval and A. Théorêt. *J. Electrochem. Soc., Solid State Sci. and Technol., 122*(4), 530 (1975).
54. S. Walker and R. F. Barrow. *Trans. Far. Soc., 50*, 541 (1954).
55. D. D. Neiswender. In *Chemical Reactions in Electrical Discharges*, ed. B. D. Blaustein. Advances in Chemistry Ser., *No. 80*. Amer. Chem. Soc.: Washington, D.C., 338 (1969).
56. B. W. Brooks and R. M. Sambrook. *J. Appl. Chem. Biotechnol., 22*, 9 (1972).
57. M. W. Ranney and W. F. O'Connor. In *Chemical Reactions in Electrical Discharges*, ed. B. D. Blaustein. Advances in Chemistry Ser., *No. 80*. Amer. Chem. Soc.: Washington, D.C., 297 (1969).
58. P. L. Kronick, K. F. Jesch, and J. E. Bloor. *J. Polym. Sci., A-1, 7*, 767 (1969).
59. A. R. Denaro, P. A. Owens, and A. Crawshaw. *Eur. Polym. J., 4*, 93 (1968).
60. J. E. Viguié and J. Spitz. *J. Electrochem. Soc., Solid State Sci. and Technol., 122*(4), 581 (1975).
61. R. L. Paciorek and R. H. Kratzer. *Can. J. Chem., 48*, 1777 (1970).
62. J. M. Brown, P. B. Coates, and B. H. Thrush. *Chem. Commun., 884* (1966).
63. J. V. Michael and N. Niki. *J. Chem. Phys., 46*, 4965 (1967).
64. H. Yasuda and T. Hirotsu. *J. Polym. Sci., Polym. Chem. Ed., 16*, 229 (1978).
65. H. Melvill and B. G. Govenlock. *Experimental Methods in Gas Reactions*. Macmillan: London (1964).
66. H. Yasuda, H. C. Marsh, M. O. Baumgarner, and N. Morosoff. *J. Appl. Polym. Sci., 19*, 2845 (1975).

67. H. Yasuda, M. O. Baumgarner, H. C. Marsh, and N. Morosoff. *J. Polym. Sci., Polym. Chem. Ed., 14,* 195 (1976).

68. N. Morosoff, W. Newton, and H. Yasuda. *J. Vac. Sci. Technol., 15*(6), 1815 (1978).

69. R. L. Cormia, K. N. Tsujimoto, and S. Anderson. U.S. Pat. 4,013,532 (1978).

70. L. F. Thompson and K. G. Mayhan. *J. Appl. Polym. Sci., 16,* 2317 (1972).

71. A. R. Westwood. *Europ. Polym. J., 7,* 363 (1971).

72. H. Kobayashi, M. Shen, and A. T. Bell. *J. Macromol. Sci., Chem., A8*(8), 1345 (1974).

73. J. M. Holovka. Scandia Labs., Rt. SLA 740181 (1974).

74. M. M. Millard, J. J. Windle, and A. E. Pavlath. *J. Appl. Polym. Sci., 17,* 2501 (1973).

75. S. Morita, T. Mizutani, and M. Leda. *Japan J. Appl. Phys., 10,* 1275 (1971).

76. J. P. Wightman and N. J. Johnston. In *Chemical Reactions in Electrical Discharges,* ed. B. D. Blaustein. Advances in Chemistry Ser., *No. 80.* Amer. Chem. Soc.: Washington, D.C., 322 (1969).

77. B. T. Clark and D. Shuttleworth. *J. Polym. Sci., Polym. Chem. Ed., 17,* 1317 (1979).

78. A. Kupperman and J. Raff. *Discuss. Farad. Soc., 35,* 30 (1963).

5. Plasma Deposition of Films

1. Introduction

Plasma-deposition processes have become an industrially recognized technique, particularly in areas that impose high demands on specific performance functions. Thus this technique has now gained wide recognition in the area of electronic circuitry protection in which the coated layer must be highly conformal and the ultra-thin film must be pinhole-free. Nonpyrolitic coating methods investigated in the recent past for this application include, in addition to glow discharge, electron-beam-activated deposition and deposition initiated through UV irradiation. The glow-discharge technique appears to have gained the most attention in recent times.

Other specific performance targets for such films include dielectrics for conversion devices; capacitors; optical windows and lenses; passivation of semiconductors; moisture-resistant antireflection; barriers to specific vapors, liquids, and solutions; and insulation and chemical resistance. One of the earlier problems in this development was the slow, uneconomic rate of polymer deposition. A great deal of effort has been devoted, particularly during the last decade, to the study of the effect of experimental parameters, carrier gases, and chemical additives. Thus electrical discharges have been investigated from d.c. to microwave frequencies for the applied electrical fields; pressures from a fraction of a torr to many atmospheres; power levels from a few watts to several megawatts; effective residence times of reactive species from a few microseconds to many seconds.

2. Plasma Parameters in Film Deposition

A. Electric Field and Gas Density

Townsend [1] and others pointed out in the forties and fifties that at low degrees of ionization and excitation of the gas, the electron-transport coefficients as well as the rate coefficients for reactions induced by electrons in a d.c. electric field E and for a given gas mixture depend essentially on the value E/N, where N is the total gas density.

The E/N value required to produce a given reaction rate coefficient varies markedly with the gas. As an example, an E/N value of only 3×10^{-18} volts-cm^2 is required to obtain a mean electron energy of 1.5 eV in pure Ar, whereas for the same mean electron energy in a highly polar gas such as H_2O an E/N value of 6×10^{-16} volts-cm^2 is needed [2]. Phelps pointed out that these differences arise from the fact that at these energies in pure Ar the electrons lose energy only through the rebound of the heavy Ar atoms in an elastic collision, while in H_2O the electrons lose energy very rapidly because of the excitation of rotational and vibrational states of the molecule. Because of the large cross sections for rotational and vibrational excitation of some molecular gases, small additions of rare gases cause large charges in the mean electron energy and in the electron-transport coefficient at fixed E/N.

The E/N parameters have been widely used to compare measurements of the rate coefficients for the ionization of atoms and molecules by electron impact. Attempts have also been made to obtain data for the correlation of experimental rate coefficients for chemical reactions such as molecular dissociation. Such correlations are generally recognized as essential for the understanding and prediction of the rates of chemical reactions in electrical discharges under a variety of experimental conditions.

B. Effective Electric Field Strength

The concept of effective electric field strength, originally applied to include the observed frequency dependence of the electric field strength required for gas breakdown, has been found to be of considerable general utility for homogeneous discharges; it makes it possible to compare the effect on the electrons of an applied electric field of a given frequency to that of an equivalent d.c. applied field. The effective d.c. electric field, E_e, is related to the r.m.s. value of the applied field, E, by

$$E_e = \frac{E}{(1 + \omega^2)/(\nu_e^2)^{\frac{1}{2}}} \qquad \text{(Eq. 5.1)}$$

where ν_e is the effective frequency of electron collisions with gas molecules, and ω is the radian frequency of the applied field. The electron collision frequency under discharge conditions [3] is 10^{-7} Nsec^{-1}, so that at a pressure of 1 torr at 300°K the quantity ω/ν_e equals 1 at a frequency of 500 MHz [2]. At considerably lower frequencies, that is, below 150 MHz, $E_e = E$, and the rate at which energy is adsorbed by electrons from the electric field is independent of the frequency. The effective field relation has been found [4] to apply to He and H_2 under gas breakdown conditions, or high E_e/N values.

The effective field concept is believed [3] to be particularly useful for the accurate prediction of quantities that vary slowly with E_e/N, that is, electron-drift velocity, average electron energy, discharge-maintenance field strengths, and certain reaction rate coefficients. The effective field relation should at least be accurate at low ω/N in gases such as Ar and CH_4, where the cross section for momentum-transfer collisions between electrons and gas molecules is a rapidly increasing function of electron energy over much of the energy range of interest [3]. In cases where the effective field concept is not sufficiently accurate, data such as rate coefficients at various ω and N in a given gas can be shown to depend upon the parameters E/N and ω/N.

C. Breakdown Parameters

Breakdown parameters have been widely investigated. Measurements of breakdown, achieved by slowly increasing d.c. voltage and using electrode gaps with a small spacing in relation to the wall distance, are generally given in terms of breakdown voltages as a function of pressure $\times$ gap spacing, d. By dividing the breakdown voltage by Nd, the results can be expressed as the E/N value required to induce breakdown for a given value of Nd. The significance of Nd results from the requirement that each electron produce a sufficient number of positive ions or excited molecules to cause the release of an additional electron from the cathode.

When the voltage is applied in the form of a short pulse of length Δt, the product $N\Delta t$ would then also represent an important parameter because the ionization must be produced within a time Δt with a rate coefficient given by E/N. In the limit of very short pulses the Nd parameter becomes unimportant [5]. At low a.c. frequencies the parameters needed to correlate breakdown data for a given gas are the same as for d.c. breakdown, except in cases where the number of electrons, available for initiating breakdown, is low [6].

The rate coefficient for ionization is often expressed as a function of E_e/N, E/N, ω/N, or N/λ, where λ is the wavelength corresponding to ω. The balance between the ionization rate and the diffusion loss introduces the parameter $N\Lambda$ [4], where Λ is the diffusion length. It has been noted [2] that the reflection coefficient for ions and electrons at surfaces is small; therefore, the diffusion losses can usually be calculated on the basis of zero concentration at the wall.

The parameters E/N, $N\lambda$, and $N\Lambda$, or combinations of these, have been employed to correlate experimental data for a given gas and a wide range of λ and N [4], including high gas densities.

D. Parameters for Discharge Maintenance

This section deals chiefly with the electric field strength required to maintain a given electron density. At moderate degrees of ionization the steady-state discharge conditions are represented by a balance between the rates of ionization by electrons and the rates of loss of electrons by diffusion to the walls or by flow of electrons and ions to electrodes. Both the ionization rate coefficient and effective electron temperature can be related to E_e/N [7] and can be used to predict the electron temperature required to maintain a d.c. positive column or r.f. plasma for various values of $N\Lambda$.

It may be noted that this is the same condition as that for the diffusion-controlled breakdown, previously discussed, with the exception that the diffusion loss rate for the electrons has been considerably reduced by space-charge effects [7]. The lower loss rate means that the ionization rate coefficient that is needed to maintain the discharge is reduced, that is, the E_e/N required is reduced. In view of the similarity of the breakdown and diffusion-controlled plasma conditions, the same parameter may be used to compare results from a given gas but different experimental conditions E/N, $N\Lambda$, and $N\lambda$.

Both the production and loss rates for a diffusion-controlled plasma can be determined by the interaction between the discharge and the electrical circuitry [8]. As a general rule, the larger the discharge current, the lower the electric field strength available for the discharge. In the case of d.c. or low-frequency a.c. discharges, a large fraction of the applied voltage is frequently needed to release the required electron current from the cathode and to satisfy space-charge relations at the cathode and anode [9].

For conditions in which the effective electron temperature is much larger than the gas temperature, the ratio of the effective diffusion coefficient in the presence of space charge, D_s, to the diffusion coefficient for free electrons, D_e, is given by Phelps [10] by

$$\ln D_s/D_e = [1 + (\lambda_D/\Lambda)^{0.76}]^{-1} \ln \mu_+/\mu_e \qquad \text{(Eq. 5.2)}$$

where μ_+ and μ_e are the mobilities of the positive ions and electrons, respectively.

Additional processes, such as ionization of excited states of molecules or atoms by electrons, and ion-electron recombination by either second- or third-order processes, may also be required to describe the discharge. When the rates of these processes are comparable to the rates of ionization by electron impact and loss by diffusion, changes in the E_0/N would be only small [11]. When the rates of these processes become large, however, they tend to constrict the discharge. Furthermore, energy-sharing collisions between electrons may also result in higher ionization rates [12]. The presence of either stationary or moving nonuniformities, that is, striations, has also been observed [11] to complicate certain low-pressure discharges.

E. Constriction of the Discharge

The phenomenon of constriction of the current channel has been observed at high gas pressures [13]. These do not generally apply to glow discharges described in this book and are discussed and referenced in greater detail by Phelps [2].

F. Excited-State Concentrations

Excited states are produced by electron-molecule collisions and are destroyed by excited molecule–ground state molecule collisions, by electron collisions, by wall collisions, or by spontaneous radiation. At sufficiently high gas densities, spontaneous radiation and loss by diffusion to the wall are not important. Therefore, for fixed E_e/N, the degree of excitation will increase with the extent of gas ionization at low degrees of ionization and will be determined by the electron temperature at high degrees of ionization.

It has been shown [3] that the rates of vibrational excitation of molecules by electron impact can be very large. These large cross sections may be responsible for the high concentrations of vibrationally excited molecules in flow systems [14]. Due to these large cross sections, a large fraction of the energy absorbed by the electrons from the electric field is transferred to the gas molecules in the form of vibrational excitation. At higher gas pressures and translational temperatures, this vibrational energy is degraded to translational energy.

3. Effect of Pressure on Deposition Rate

A. Introduction

An important conclusion derived from the plasma kinetic theory (chapter 4, section 2B) is that the average electron energy in terms of the electron temperature, T_e, is solely a function of E/p, that is, the electric field strength, E, at the gas pressure, p. Many workers have also used the parameter E/N, where N is the gas density, to compare measurements of the rate coefficients for ionization of atoms and molecules by electron impact. For each E/p, the effects of the extent of ionization are identified by n/N, where n is the electron (or ion) density.

If we recall reaction 29 in chapter 4, the rate of atomic hydrogen formation in ethylene initiation may be expressed by:

$$r_2 = k_2[e] \, [C_2H_4] \qquad \text{(Eq. 5.3)}$$

where [e] and $[C_2H_4]$ are the concentrations of electrons and ethylene, respectively. The reaction rate constant, k_2, is dependent upon the average electron energy, which is characterized by the aforementioned parameter E/p, as measured in volts/cm torr, where here it represents the ratio of the electric field strength, required to sustain the discharge, to the gas pressure [15]. An example of the relationship between E/p and k_2 is illustrated in figure 5.1 [7].

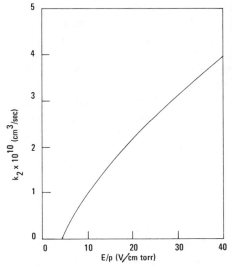

Figure 5.1. Relationship between the reaction rate constant, k_2 (equation 5.3), and E/P. *Source:* Adapted from H. Kobayashi, M. Shen, and A. T. Bell, Res. Contr., AD-762, 480, Office of Naval Res. (June 15, 1973).

It has been shown from discharge physics [15] that E/p is related to $p\Lambda$, where $\Lambda = d/\pi$ and d is the electrode separation. This latter parameter has also been frequently used in the literature to relate deposition rates. Furthermore, the quantity $[e]/\bar{P}d$, in which $\bar{P}$ is the power density, has also been expressed as a function of $p\Lambda$. Thus, once p, d, and $\bar{P}$ are specified, k_2 and $[e]$ can be determined [16].

Because of the large number of parameters affected, the effect of pressure on the deposition rate is most difficult to interpret. As the pressure increases, E/p decreases, thus causing a reduction in the value of k_2. The concentration of electrons, $[e]$, is also reduced. While decreases in both k_2 and $[e]$ operate to reduce the rate of atomic hydrogen production, they are partly offset by the increase in ethylene concentration. Analogy with other systems [17,18] indicates that r_2 will increase, thus causing the free-radical production to rise. The increased rate of the free-radical formation is opposed by processes that consume free radicals and are also enhanced by increases in monomer pressure (ethylene). As a result of these two processes, it is assumed that the free-radical concentration will increase with pressure but at a decreasing rate.

At the same time, the concentration of monomer on those surfaces where a film is deposited will also increase with rising monomer pressure. Since on a surface the number of adsorption sites is fixed, the surface concentration of adsorbed monomer will approach a maximum equivalent to monolayer coverage.

The dependence of the rates of polymer deposition on pressure can now be explained in terms of the anticipated responses of gas-phase free-radical concentration and the surface concentrations of adsorbed monomer. As the pressure is increased from very low values, the free-radical concentration and the monomer concentration, as well as the concentration of adsorbed monomer, all increase. Over a low range of pressure the deposition rates are expected to increase. When the pressure is raised further, the gas-phase polymerization process will begin to decline as a result of chain termination reactions that are second order in free-radical concentration [16]. The rate of heterogeneous polymerization will approach a maximum because of a saturation on the surface with adsorbed monomer. Yasuda and Lamaze [19,20] have demonstrated that the pressure of a steady-state flow in plasma, p_g, is different from the pressure of steady-state flow of monomer before plasma is initiated, p_0, and that the ratio $\delta = p_g/p_0$ can be correlated to the characteristic behavior of an organic compound in plasma.

In a closed system, the original pressure, p_0, changes to p_g within a few seconds in most cases; the value of $\gamma° = p_g/p_0$ is then fairly close to that of the hydrogen yield, that is, the number of hydrogen molecules evolved from a monomer when polymerization takes place. The value of $\gamma° = p_g/p_0$ is generally considered an important characteristic of a monomer and can be readily measured. The value of δ in a steady flow system is generally not identical with that of $\gamma°$ and depends on other factors in plasma polymerization.

If the relationship between p_g (in a steady-state flow) and $\gamma°$ (in a closed system) is known, the effect of operational factors such as flow rate, discharge power, and so forth on the characteristic nature of plasma deposition can be monitored by noting p_g in a steady-state flow experiment. It should be noted that the system pressure, p_0, is highly dependent on the pumping rate rather than a unique function of the flow rate. The system pressure, p_0, before the discharge is initiated, depends on both monomer flow rate, F_0, and pumping rate of the system. For instance, with a liquid nitrogen trap that acts as an excellent pump for condensible vapors, the system pressure is considerably lower compared with that without such a trap, even if the monomer flow rate is identical. The relationships between p_g and p_0, and p_g and F_0 [21]

$$\log p_g = \log c + d \log p_0 \qquad \text{(Eq. 5.4)}$$

and

$$\log p_g = \log c' + d' \log F_0 \qquad \text{(Eq. 5.5)}$$

(where C is a parameter related to γ and d a characteristic constant) have been shown [21] to apply in the case of ethylene plasma discharge (figure 5.2). The values of parameters c, d, c', and d', as shown in equations 5.4 and 5.5, are summarized in table 5.1.

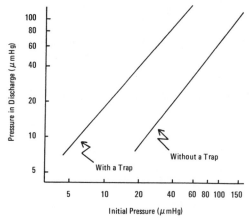

Figure 5.2. Pressure in discharge versus initial pressure, p_0, for ethylene. *Source:* Adapted from H. Yasuda and T. Hirotsu, *J. Appl. Polym. Sci., 22,* 1195 (1978). Copyright © 1978, John Wiley & Sons, Inc. Reprinted by permission of John Wiley & Sons, Inc.

Table 5.1. Dependence of p_g on flow rate and initial pressure (equations 5.4 and 5.5)

Monomer	With trap	Without trap
Acetylene	$c = 0.56$	$c = 0.064$
	$d = 1.03$	$d = 1.20$
	$c' = 4.0$	$c' = 4.5$
	$d' = 0.68$	$d' = 0.93$
Ethylene	$c = 1.29$	$c = 0.179$
	$d = 1.14$	$d = 1.28$
	$c' = 24$	$c' = 24$
	$d' = .0.69$	$d' = 0.73$
Acrylonitrile	$c = 0.220$	$c = 0.119$
	$d = 1.29$	$d = 1.22$
	$c' = 4.0$	$c' = 10$
	$d' = 0.95$	$d' = 0.74$

Source: H. Yasuda and T. Hirotsu, *J. Appl. Polym. Sci., 22,* 1195 (1978). Copyright © 1978, John Wiley & Sons, Inc. Reprinted by permission of John Wiley & Sons, Inc.

B. Experimental Data

The effect of pressure on the ethylene flow rate is illustrated in figure 5.3.

It is shown that, at low pressures and low monomer flow rates, the polymer is deposited both as a powder and as a film. At high pressures and high flow rates, the deposition appears to be in the form of an oily film. Only at low pressures and high flow rates is a solid, pinhole-free film deposited. If the pressure in the reactor is high, and the flow rate is low, the discharge becomes unstable. Under the same conditions, decreased power makes it possible to produce a film at lower flow rates. This trend has been confirmed for powers

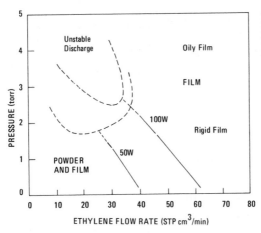

Figure 5.3. Effect of pressure on flow rate. *Source:* Adapted from H. Kobayashi, M. Shen, and A. T. Bell, Res. Contr. AD–762, 480, Office of Naval Res. (15 June 1973).

of 25 to 150 watts. It is noted that the transitions from powder to film and film to oil regions are gradual and that the demarcation lines in figure 5.3 should be regarded as approximate values.

The general observation that above a certain minimum pressure, the rate of deposition is essentially independent of pressure [22] (figure 5.4) is taken as an indication of a surface reaction. In cases where a gas-phase polymerization occurs, a proportionality to pressure is observed. The adsorption of a monomer at constant temperature on an electrode has earlier been expressed [22] by the relation

$$x/m = \int (p/p_0)T \qquad \text{(Eq. 5.6)}$$

where x/m is the amount adsorbed and p/p_0 is the relative vapor pressure, that is, the vapor pressure at the operating temperature. The rate of deposition of a monomer at a fixed current density is dependent on the number of molecules or active species adsorbed on the electrode. At low pressures, that is, $p/p_0 \rightarrow$

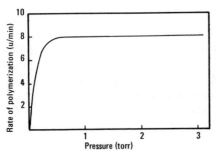

Figure 5.4. Effect of pressure on deposition rate of polystyrene. *Source:* Adapted from T. Williams and M. W. Hayes, *Nature*, London, *209*, 769 (1966). Reprinted by permission from *Nature*. Copyright © 1966 Macmillan Journals Limited.

0, this number is small. If there are enough active species striking the surface to permit every adsorbed molecule to react, then the reaction is proportional to the number of adsorbed molecules; p and p_0 are constant. Above a certain saturation pressure, the number of adsorbed molecules becomes so large that the number of active species formed is now dependent on the number of excited species striking the surface. Under these conditions, the rate of deposition is independent of pressure and is dependent only on the current density. The shape of the curve in figure 5.4 is somewhat similar to a normal adsorption isotherm, which is to be expected in light of the preceding discussion. The effect of pressure on deposition rate for vinyl chloride is shown in figure 5.5 for two fixed current densities (3.14 MHz r.f.).

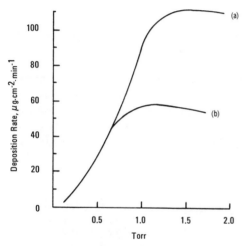

Figure 5.5. Deposition of polymer in a glow-discharge monomer vinyl chloride: (a) 1.76 mA cm^{-1}; (b) 1.40 mA cm^{-1}; temperature 15°C. *Source:* Adapted from A. R. Westwood, *Europ. Polym. J.*, 7, 363 (1971). Reprinted with permission from *Europ. Polym. J.* Copyright 1971, Pergamon Press, Ltd.

Table 5.2 shows the rates of deposition from a number of monomers, averaged with respect to time and with the discharge parameters indicated. These data should only be considered as approximate values, since the variation of polymer density with changing stoichiometry was not taken into account [23]. The data do not represent maximum rates of deposition.

The parameter largely controlling the degree of adsorption at a fixed temperature is the ratio of the gas pressure, p, to the saturated vapor pressure, p_0. At a fixed pressure, which therefore implies a fixed number of adsorbed species, the polymer deposition rate is found to increase with increased current density; this proceeds to a limiting value above which the rate is no longer current-dependent [23]. At a higher pressure, the level of the limiting current is observed to increase.

At a fixed current density, the deposition rate is pressure-limited at low pressures, but it becomes current-limited at higher pressures. The higher the

Table 5.2. Rates of deposition of plasma polymers

	Discharge conditions			
Monomer	Current (mA/cm²)	Pressure (torr)	Temp. (°C)	Rate (Å/sec)
Vinyl acetate	2.43	0.99	0	40
Methyl methacrylate	2.43	1.96	0	130
Vinyl chloride	1.53	0.99	0	160
Vinyl chloride	3.79	1.01	20	220
Vinylidene fluoride	3.73	0.82	20	260
Vinyl fluoride	2.01	0.63	0	85
Styrene	1.28	0.98	20	40
Tetrafluoroethylene	2.42	0.95	20	4
Ethylene	2.01	0.63	20	14
Vinylidene chloride	2.43	0.63	20	86

Source: A. R. Westwood, *Europ. Polym. J.*, *7*, 363 (1971). Reprinted with permission from *European Polymer Journal.* Copyright © 1971, Pergamon Press, Ltd.

current density, the higher the pressure at which this current limitation appears [23]. From the above discussion, it appears interesting to plot the rate of polymer deposition as a function of relative pressure, p/p_0. For methyl-methacrylate, the plot is shown in figure 5.6. It is noted that these data were obtained at constant p, with a varying p_0, while in the case of an adsorption isotherm, p is varied.

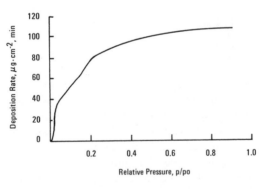

Figure 5.6. Effect of relative pressure, p/p_0, on deposition rate of methylmethacrylate; 1.96 torr. *Source:* Adapted from A. R. Westwood, *Europ. Polym. J.*, *7*, 363 (1971). Reprinted with permission from *Europ. Polym. J.* Copyright 1971, Pergamon Press, Ltd.

It has frequently been observed that polymers from monomers such as styrene and methylmethacrylate are soft and tacky. Under such conditions, which correspond to the multilayer region of the adsorption isotherm, loosely bound or unreacted monomer molecules presumably become trapped in the rapidly growing polymer. Conversely, polymers deposited at low pressure, implying low surface coverage with adsorbed material, are generally harder and more tightly adherent films. Therefore, where coatings are concerned, it does not appear possible to take advantage of the higher rates of polymer deposition observed under high coverage conditions without losing some of the desirable coating properties inherent in plasma depositions. Other techniques have been used, such as selected carrier gas and coaddition of catalytically acting molecules, to enhance deposition rates without loss in coating quality.

The adsorption behavior of the monomer may be represented by the BET equation for multilayer adsorption:

$$\frac{p}{v(p_0 - p)} = \frac{1}{v_m C} + \frac{(C - 1)}{v_m C} \frac{p}{p_0} \qquad \text{(Eq. 5.7)}$$

where v = volume of gas adsorbed, v_m = volume of a monolayer, p = pressure in the system, p_0 = the saturated vapor pressure of the adsorbate at the temperature of the experiment, and C = constant.

The BET equation is in good agreement with experimental observations up to relative pressures, p/p_0, of about 0.4. The shape of the BET curve is largely governed by the value of the constant C, which is given approximately by

$$C \simeq \exp (\Delta H_L - \Delta H_1)/RT \qquad \text{(Eq. 5.8)}$$

where ΔH_L = the heat of liquification of the adsorbate and ΔH_1 = the heat of adsorption of the first molecular layer.

With C values between 1 and 10, the BET curve is approximately linear up to relative pressures of 0.5 [24]; therefore, within these limits, the concentration of adsorbed monomer can be taken as proportional to the pressure. From the BET equation, it can be seen that with $C = 1$, the volume of gas adsorbed is not equal to the monolayer volume until the relative pressure is 0.5.

In r.f. flow discharges of styrene, recent experiments [25] were carried out at room temperature and 18°C; the styrene vapor pressure was 4.8 torr. The maximum pressure of styrene used in the experiments was 2 torr, which corresponds to a relative pressure of 0.4. In this case, where the adsorption of styrene onto polystyrene is considered, ΔH_L and ΔH_1 are considered to be close to equal, and the approximate value of C will be unity. Therefore, under these experimental conditions, the surface of the electrode will not be completely covered, and a radical site may be formed at a point on the surface where no adsorbed monomer is adjacent. Such a radical can participate in a propagation reaction only when monomer diffuses to the site from the gas phase. The rate of monomer arrival at the electrode is proportional to its pressure. On the basis of these arguments, the rate of the propagation step will be proportional to monomer pressure regardless of whether a radical reacts with adsorbed monomer or with monomer diffusing from the gas phase.

In their experiments, the investigators confirmed experimentally the general slope for styrene from pressure versus deposition rates as shown in figure 5.4. Moreover, they found that the radical production, r, is a process of both first and second order in power,

$$r = k''W^2 + k'W \qquad \text{(Eq. 5.9)}$$

and may be written for the rate of deposition:

$$\text{Rate} = r \; \frac{p}{p + A + (A^2 + Br)^{\frac{1}{2}}} \qquad \text{(Eq. 5.10)}$$

Combination of these two equations yielded:

$$\frac{\text{Rate of deposition}}{\text{of polymer}} = \frac{p(k''W^2 + k'W)}{p + A + [A^2 + B(k''W^2 + k'W)]^{\frac{1}{2}}} \qquad \text{(Eq. 5.11)}$$

The effect of pressure of 2-vinylpyridine under 10 KHz a.c. discharge on the deposition rate, R, was recently studied [26] over a pressure range between 40 to 280 millitorr and at a constant current density of 0.1mA/cm^2. The results are illustrated in figure 5.7. Within the pressure range of 80 to 180 millitorr, R is proportional to monomer pressure. Below 80 millitorr, the discharge is unstable, as evidenced by the potential required to maintain the current density at 0.1mA/cm^2. Above 180 millitorr, R begins to fall off concurrently with the appearance of "fog," which was observed to emanate from the glow region near the electrodes. Most of this material was carried by the gas stream into the vacuum pump. Evidently, the appearance of the fog was caused by vapor-phase polymerization.

In an interesting experiment Halovka [26] investigated the thickness of adsorbed monomer as a function of monomer pressure. The dependence of $N_{2\text{-VP}}$, the number of monomer molecules adsorbed per unit area, on $P_{2\text{-VP}}$ could not be determined, since the activated surface of PVP during the deposition process is difficult to reproduce in static experiments. Instead, PVP was deposited onto aluminized quartz crystals, and a thickness monitor [27] was used to determine the amount of 2-vinyl pyridine adsorbed on PVP as a function of 2-VP monomer pressure, argon pressure, and temperature.

As a result, $N_{2\text{-VP}}$ was found to be directly proportional to $P_{2\text{-VP}} > 80$ millitorr (figure 5.8), and inversely proportional to the substrate temperature.

Figure 5.7. Effect of varying the pressure of 2-vinyl pyridine on deposition rate. *Source:* Adapted from J. M. Halovka, Sandia Labs., Res. Rept. No. SLA 74–0181 (1974).

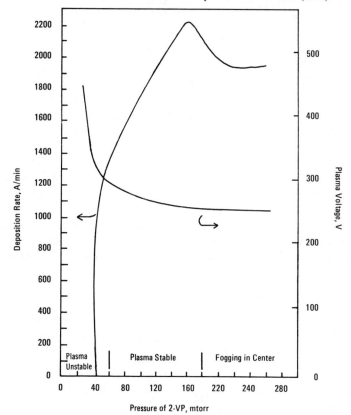

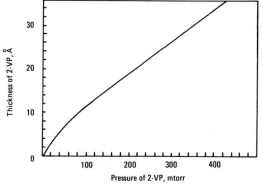

Figure 5.8. Adsorption isotherm for 2-vinyl pyridine on polyvinyl-pyridine. *Source:* Adapted from J. M. Halovka, Sandia Labs., Res. Rept. No. SLA 74–0181 (1974).

In addition, $N_{2\text{-VP}}$ was independent of the argon pressure. The overall expression for $N_{2\text{-VP}}$ is given by

$$N_{2\text{-VP}} = \frac{P_{2\text{-VP}} + k_3}{T} \qquad \text{(Eq. 5.12)}$$

for $P_{2\text{-VP}} > 80$ millitorr.

Halovka's [26] speculation as to the nature of the species that initiates the polymerization is interesting. He argues as follows: the electrode surface is subject to impinging ions, electrons, and protons, and all are possible initiators. In a direct-current discharge, polymer was found only on the cathode, which suggests that ions are more effective than photons or electrons in initiating polymerization. The nitrogen in the pyridine ring would tend to form a complex with the ion, however, thus reducing its reactivity. A more likely possibility is that the impinging ion strikes the surface and generates free-radical sites on the surface. Those, in turn, initiate polymerization or recombine with other free-radical sites to form polymer crosslinks. It must be concluded that the surface-addition polymerization mechanism predominates.

4. Effect of Flow Rate

For a cylindrical tube of diameter d and length L, the number of molecules, F, flowing from side 2 (pressure 2 and temperature 2) to side 1 (pressure 1 and temperature 1) is given by [28]:

$$F = \frac{Nd^3\pi}{3L\,(2\,\pi\,MR)^{\frac{1}{2}}} \left[\frac{P_2}{\sqrt{T_2}} - \frac{P_1}{\sqrt{T_1}} \right] \qquad \text{(Eq. 5.13)}$$

where M = molecular weight, N = Avogadro's number, and R = gas constant. The flow is linearly proportional to the pressure drop, $p_2 - p_1$, if $T_2 \simeq T_1$.

In this case the Poiseuille relation for the flow of liquid through a tube is applicable [28]. If flow is isothermal, as in a short tube, the amount of gas in moles sec^{-1} flowing through a tube of length L cm is given by

$$F = \frac{(d/2)^4 \cdot \pi\,(p_2{}^2 - p_1{}^2)}{16\,L \cdot \eta \cdot R \cdot T} \qquad \text{(Eq. 5.14)}$$

where p_2 and p_1 are pressures (dynes/cm^2) at the two ends of the tube, R is the gas constant, and η is the viscosity of the gas. When p_1 is maintained at zero, as compared with p_2, flow in this event is proportional to the square of the pressure, p_2.

In most practical cases, the reaction vessel is not a straight cylindrical tube, and the pressure is not measured at the inlet and outlet of the gas flow. Therefore, in the pressure range most frequently employed in glow discharges, and at $p_1 \simeq 0$, the gas flow, F, is empirically given by [21]

$$F = ap^b$$ (Eq. 5.15)

where the value of b lies between 1 and 2 and p is the system pressure of the reaction vessel measured at a fixed point.

Both the flow rate, F, and the pressure, p, of the flow system of a gas are controlled by input rate and the pumping rate. In addition to the pump efficiency, a low-temperature gas trap can also act as an excellent pump for condensible vapors, provided that the trap is situated sufficiently close to the outlet of the reaction vessel.

Experiments conducted by Yasuda and his coworkers [29] have shown that the pressure of a steady-state flow in plasma, p_g, is different from the pressure of the steady-state flow of monomer, p_o, before plasma is initiated. In a closed system, the original pressure, p_o, changes to p_g fairly rapidly, and the value of $\gamma° = p_g/p_o$ is close to the value of hydrogen yield for many hydrocarbons, amines, and nitriles [29].

In a closed system, the total number of monomer, n, changes to γn. In a flow system, plasma polymerization can be considered a process that changes the flow rate of gas

$$dn/dt \rightarrow \gamma \, dn/dt$$ (Eq. 5.16)

The flow rate, F_1, at which monomer enters the reaction vessel, is given by the pressure of the flow system, p_1, of the monomer before discharge, according to equation 5.15.

$$dn/dt = F_1 = a_1 p_1{}^{b_1}$$ (Eq. 5.17)

As plasma polymerization commences, the total number of molecules in gas phase changes to $\gamma \, dn/dt$. This is the amount of gas that would have to be pumped out to maintain a steady-state flow.

$$F_2 = a_2 p_2{}^{b_2}$$ (Eq. 5.18)

Hence,

$$a_2 p_2{}^{b_2} = a_1 p_1{}^{b_1}$$ (Eq. 5.19)

or

$$a_2 p_2{}^{b_2} = \gamma \, F_1$$ (Eq. 5.20)

Subscript 1 denotes the monomer, and subscript 2 denotes the gas to be pumped from the system.

The relationship between monomer flow rate and system pressure as represented by equation 5.15 is illustrated in figure 5.9 on ethylene monomer. It is noted that the system pressure, p_0, prior to plasma initiation, is dependent on both monomer flow rate, F_0, and the pumping rate of the system. Thus, the system pressure, p_0, is not a unique function of the flow rate but is greatly dependent on the system's pumping rate.

From the above equations, the pressure in discharge, p_g, started from a steady-state flow at p_0, is given by

$$\log p_g = \log c + d \log p_0 \qquad \text{(Eq. 5.21)}$$

where

$$c = (\gamma \, a_1/a_2)^{1/b_2} \qquad \text{(Eq. 5.22)}$$

$$d = b_1/b_2 \qquad \text{(Eq. 5.23)}$$

or

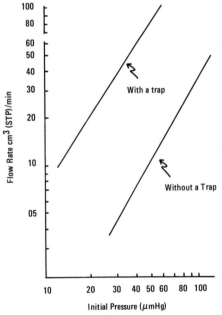

Figure 5.9. Flow rate versus pressure for ethylene with and without a liquid nitrogen trap. *Source:* Adapted from H. Yasuda and T. Hirotsu, *J. Appl. Polym. Sci.*, 22, 1195 (1978). Copyright © 1978, John Wiley & Sons, Inc. Reprinted by permission of John Wiley & Sons, Inc.

$$\log p_g = \log c' + d' \log F_0 \qquad \text{(Eq. 5.24)}$$

where

$$c' = (\gamma \, a_2)^{1/b_2} \qquad \text{(Eq. 5.25)}$$

$$d' = 1/b_2 \qquad \text{(Eq. 5.26)}$$

The validity of equations 5.21 and 5.24 can be examined if the product gases and parameters a_2 and b_2 are known. The flow characteristics of monomers and hydrogen are summarized in table 5.3 [21].

This flow rate in a noncompressive system, such as a liquid, has two interpretations, namely, the input rate of mass and the "sweeping rate" of the molecules, from which the residence time of the molecules can be estimated. When applied to a "flow" of gas, however, the second process should not generally be considered as similar to the flow of a liquid [30]. In a gas flow under vacuum, the absolute velocity of gas molecules as well as the diffusional displacement velocity (Brownian motion) are large. The gas flow rate represents not the velocity of the individual gas molecules but only the total flux. Therefore, it does not represent the "sweeping rate" of individual molecules. Thus, for a reactive plasma, flow rate is to be taken as input rate only, that is, in the nonplasma state, since knowledge of the plasma gas phase is generally not available.

Furthermore, the formation of active species leading to deposition occurs primarily in the glow region, the volume, V_g, of which is not always the same

Table 5.3. Flow characteristics of monomers and hydrogen

Gas	With trap	Without trap
Hydrogen	$a_2 = 1.23 \times 10^{-2}$ $b_2 = 1.32$	$a_2 = 1.00 \times 10^{-2}$ $b_2 = 1.30$
Acetylene	$a_1 = 4.88 \times 10^{-2}$ $b_1 = 1.35$	$a_1 = 0.34 \times 10^{-2}$ $b_1 = 1.50$
Ethylene	$a_1 = 2.56 \times 10^{-2}$ $b_1 = 1.50$	$a_1 = 0.189 \times 10^{-2}$ $b_1 = 1.64$
Acrylonitrile	$a_1 = 12.3 \times 10^{-2}$ $b_1 = 1.73$	$a_1 = 0.11 \times 10^{-2}$ $b_1 = 1.79$

Note: $F = ap^b$, F is the flow rate in cm^3 (STP)/min, and p is the flow pressure in μm Hg. H_2 is considered as the product gas, and subscript 2 is used for H_2.

Source: H. Yasuda and T. Hirotsu, *J. Appl. Polym. Sci.*, *22*, 1195 (1978). Copyright © 1978, John Wiley & Sons, Inc. Reprinted by permission of John Wiley & Sons, Inc.

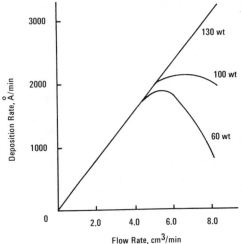

Figure 5.10. Deposition rate of acrylonitrile versus flow rate for inductive discharge. *Source:* Adapted from H. Yasuda and T. Hirotsu, *J. Polym. Sci., Polym. Chem. Ed., 16,* 743 (1978). Copyright © 1978, John Wiley & Sons, Inc. Reprinted by permission of John Wiley & Sons, Inc.

as that of the whole reactor, V_r. Flow rate is generally determined as the feed-in rate into V_r and not into V_g. For instance, in a bell-jar-type reactor, V_g is only a fraction of V_r. Therefore, the bypass ratio of flow, V_r/V_g, should be considered. This ratio V_r/V_g, however, is a variable under conditions of plasma polymerization, since effective plasma polymerization, that is, low γ value, acts as a pump and pulls the monomer into V_g from its surrounding volume [30].

The deposition rate of acrylonitrile under full r.f. glow condition is shown in figure 5.10 at three levels of discharge power. It is seen that except for the high power, the deposition rate increases until a maximum is reached, then the rate declines, depending on the discharge power. Vinyl chloride shows a deposition rate dependence on flow similar to that of acrylonitrile (figure 5.11). As the flow rate increases, the rate of polymer deposition increases

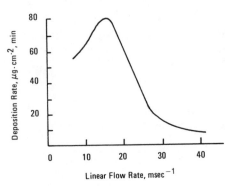

Figure 5.11. Deposition rate of vinyl chloride as a function of linear flow rate. Temperature 20°C, 200 V cm⁻¹. *Source:* Adapted from A. R. Westwood, *Europ. Polym. J., 7,* 363 (1971). Reprinted with permission from *Europ. Polym. J.* Copyright 1971, Pergamon Press, Ltd.

until a maximum is reached, after which the deposition rate declines. Westwood [31] suggests that during the first phase, the breakdown products are being swept out of the discharge zone, while the falling deposition rate with increasing flow is caused by the removal from the plasma of species capable of initiation. Under these conditions, polymer has been observed to form in the vacuum system downstream from the plasma region. Like acrylonitrile and vinyl chloride, methylmethacrylate showed an increase in deposition rate with increasing flow rate in a capacitively coupled r.f. discharge [32] to a maximum, above which the rate decreased (figure 5.12).

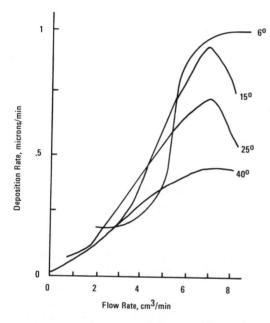

Figure 5.12. Effect of flow rate of methylmethacrylate on deposition rate. Pressure 1.9 torr, power 1.73 V-amp/cm³ at 3.14 MHz. *Source:* Adapted from K. C. Brown, *Europ. Polym. J.*, 8, 117 (1972). Reprinted with permission from *Europ. Polym. J.* Copyright 1972, Pergamon Press, Ltd.

With a change of monomer flow rate, the appearance of flow patterns in the polymer film changed. At low flows, the most concentrated deposition area was around the gas inlet end. At flows past the maximum of the curve, the polymer was concentrated around the gas exit. At maximum deposition, polymer was uniform over the electrode surfaces [32]. The chemical structure of the relationship between monomer flow rate and deposition rate was recently shown on a series of nitriles [32] (figure 5.13). Evidently, the nitriles can be placed in two groups. Those containing ethylenic unsaturation showed a marked dependence of deposition rate, while the saturated nitrile compounds displayed little increase in the rate of deposition with increased flow rate.

It should be noted that the relationships between flow rate and deposition

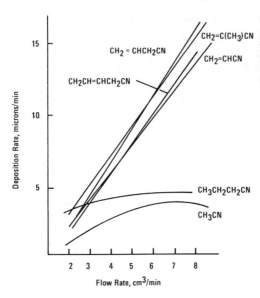

Figure 5.13. Effect of flow rate of nitrile compounds on deposition rate. *Source:* Adapted from K. C. Brown, *Europ. Polym. J.*, *8*, 117 (1972). Reprinted with permission from *Europ. Polym. J.* Copyright 1972 Pergamon Press, Ltd.

rate can be greatly affected by monomer decomposition or fragmentation. For instance, acrylic acid is known to lose carboxylic acid groups during plasma polymerization, and the polymer becomes rather hydrophobic. Investigation of the monomer content of the effluent gases revealed recently [32] that the longer the monomer spent in the discharge zone, the more of it decomposed and the less of it emerged unchanged (table 5.4) [33].

Polymer deposition rates in a glow discharge constitute an empirical parameter to express the rate by which deposition takes place under certain conditions. The rate of polymer deposition is found to be linearly proportional to the monomer feed-in rate. In an arrangement where all monomer flows uniformly through a cylindrical reactor, the rate of polymer deposition, R (g/cm²-min) is related to the weight-basis monomer flow rate, F_w, (g/min) as:

$$R = kF_w \qquad \text{(Eq. 5.27)}$$

where k = a rate constant (cm⁻²). A comparison of polymer deposition rates for vinyl monomers and their saturated homologs is shown in table 5.5 [33]. The k values were obtained by selecting the minimum wattage required to sustain a fully developed glow discharge (which, in this case, covered the entire volume of the reactor) at highest vapor pressure employed (100 μm Hg) for each monomer. The flow rate of a monomer was measured from the initial rate of pressure increase when the pumping is stopped from a steady-state flow system.

Table 5.4. Decomposition of monomer in glow discharge (in percent)

Stage	Flow[a]		
	2	6.5	9
	Allylamine		
Polymerized·	25.0	34.8	13.6
Trapped	8.9	25.0	56.1
Decomposed or lost	66.1	40.2	30.3
	Methyl methacrylate		
Polymerized	32.0	21.0	10.9
Trapped	21.0	25.6	41.8
Decomposed or lost	47.0	53.4	47.3

Note: Pressure = 1.9 torr, temperature = 15°C, power = 1.15 W/cm^2 at 13.56 MHz.

[a]Flow in units of cm^3/min.

Source: H. Yasuda, M. O. Baumgartner, H. C. Marsh, and N. Morosoff, *J. Polym. Sci., Polym. Chem. Ed., 14,* 195 (1976). Copyright © 1976, John Wiley & Sons, Inc. Reprinted by permission of John Wiley & Sons, Inc.

It should be noted that k is not a rate constant of the chemical reaction in a strict sense, since the value of k depends on geometrical factors, such as surface/volume ratio of the reactor. Nevertheless, k is a useful parameter for comparing reactivities of organic compounds in glow-discharge polymerization. The polymer deposition rate correlated to the volume flow rate in an unspecified reactor can be used only as an empirical parameter to describe the

Table 5.5. Comparison of polymer deposition rates for vinyl monomers and saturated vinyl compounds

Vinyl monomers		Saturated vinyl monomers	
Compound	$k \times 10^4$, cm^{-2}	Compound	$k \times 10^4$, cm^{-2}
4-Vinylpyridine	7.59	4-Ethylpyridine	4.72
-Methylstyrene	5.33	Cumene	4.05
Styrene	5.65	Ethylbenzene	4.52
N-Vinylpyrrolidone	7.75	N-Ethylpyrrolidone	3.76
Acrylonitrile	5.71	Propionitrile	4.49
Vinylidene chloride	5.47	1,1′ Dichloroethane	2.98
Allyl amine	2.86	n-Butyl amine	2.52
Methyl acrylate	.99	Methylpropionate	.57

Source: H. Yasuda, M. O. Baumgartner, H. C. Marsh, and N. Morosoff, *J. Polym. Sci., Polym. Chem. Ed., 14,* 195 (1976). Copyright © 1976, John Wiley & Sons, Inc. Reprinted by permission of John Wiley & Sons, Inc.

polymer deposition rate for a given set of conditions. If the reaction occurs in a small volume of the reactor, such as is the case in several bell-jar-type reactors with electrodes, the overall flow rate does not relate to the actual flow rate in the reaction volume. Therefore, correlation of polymer deposition rate with overall flow rate as a measure of monomer reactivity would be meaningless under these conditions. The k value in equation 5.27 is based on the weight-basis flow rate. The flow rate based on the volume flow rate is proportional to the numbers of molecules but not the total weight of molecules.

It has been shown before that depending on the experimental plasma conditions and sample position in the reactor, the mechanism of deposition may be simultaneous polymerization or plasma-induced polymerization. Furthermore, by conventional definition, saturated compounds have not been considered as monomers, since they would not readily provide the "active" sites for polymerization. Nevertheless, the polymer deposition rates of the vinyl monomers and their saturated counterparts are of the same order of magnitude. This is taken as an indication that plasma polymerization rather than plasma-induced polymerization is responsible for a major portion of polymer formation [33].

The deposition rates of polymers from ethane, propane, and butane were earlier (chapter 4) shown to decrease in the order $C_2H_6 > C_3H_8 > C_4H_{10}$. For these monomers, the deposition rates also decreased with increasing flow rate. The interpretation is that the residence times of the reactive species were reduced by the enhanced flow rate and that some of the reactive species were swept away before polymerization could take place. The characteristic map for plasma polymerization of ethane (figure 4.5, chapter 4) illustrates that depending on flow rate and pressure, either powder or film is formed. It has also been shown [16] that the form of polymer produced during the plasma polymerization of ethylene depends on flow rate as well as on pressure and power supplied to the plasma. At low monomer flow rates and low pressures, both powder and film are formed. At high flow rates and pressures, the polymer appears in the form of an oily film. Only at low pressure and high flow rate was a solid, pinhole-free film obtained. If the flow rate was low and the pressure in the reactor was high, the discharge became unstable.

Bell and his coworkers [16] deduced the following trends from their experimental data on ethylene:

> The rate of polymer deposition at a given monomer flow rate increases with increasing pressure and finally approaches a plateau. For a given pressure, the deposition rate first increases with increasing flow rate, reaches a maximum at an intermediate flow rate, and then decreases with further increases in flow rate.

The overall yield of polymer is reduced rapidly with increasing flow rate and more slowly with decreasing pressure. The amount of powder generated is always less than half the total yield, even at low pressure and flow rate. The other part of the product is film. At higher pressures and flow rates, there is no visible production of powder.

The chemical composition of the polymer indicates that the value of n in the formula C_2H_n is between 2.6 and 3.0. The higher values of n correspond to polymers formed at low flow rates. For a given flow rate and pressure, the chemical compositions of the powder and the film show no significant difference. In all instances, the polymer is hydrogen deficient by comparison with conventional polyethylene.

Polymers produced at low pressures are insoluble in acetone and m-xylene, indicating a crosslinked structure. Oily polymers, formed at high flow rates as well as high pressures, are either soluble or readily swollen in these solvents, leading to the conclusion that the number of crosslinks is insufficient to form a gel structure and that the polymer is most likely composed of branched oligomers.

The role of halogens in the plasma polymerization of hydrocarbons has been investigated in greater detail by Bell and his coworkers [34]. Figure 5.14 illustrates the deposition rates obtained at 0.5 torr and a power of 50 W. The order of polymer deposition rates is:

$$C_2H_3Cl > C_2H_2 > C_2H_3F > C_2H_4$$

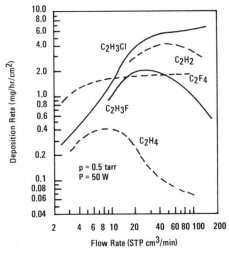

Figure 5.14. Polymer deposition rates on a weight basis as a function of flow rate. *Source:* Adapted from H. Kobayashi, M. Shen, and A. T. Bell, *J. Macromol. Sci., Chem., A8*(8), 1345 (1974). Reprinted by courtesy of Marcel Dekker, Inc., N.Y.

If the data are plotted in terms of percentage of monomer fed converted to polymer, the order of C_2H_3Cl and C_2H_2 is reversed. Characteristically, the rate of polymer deposition passes through a maximum, with increasing monomer flow rate, and the flow rate, which produces a maximum deposition rate, is shifted upscale with increasing ease of polymerization.

The form of the polymer obtained is highly dependent upon both the type of monomer and monomer flow rate. While ethylene forms a powder and a film at low flow rates and a film only at high flow rates, acetylene produces only a powder under any of the experimental conditions. Vinyl chloride yields a film as well as a small amount of powder at low flow rates and powder only at high flow rates. Vinyl fluoride behaves similarly except that the rate of powder formation is markedly reduced at higher flow rates.

Tetrafluoroethylene (TFE) polymerizes at a greater rate than ethylene (figure 5.14), but if the rates are expressed in terms of the fraction of monomer converted to polymer (figure 5.15), they are nearly comparable except at higher flow rates, where TFE polymerizes much more rapidly.

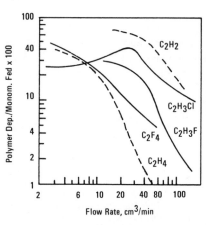

Figure 5.15. Polymer deposition rates as percent of monomer fed converted to polymer deposited as a function of flow rate. *Source:* Adapted from H. Kobayashi, M. Shen, and A. T. Bell, *J. Macromol. Sci., Chem., A8*(8), 1345 (1974). Reprinted by courtesy of Marcel Dekker, Inc., N.Y.

In contrast to monomers such as ethylene, propylene, vinyl chloride, and vinyl fluoride, which form both powder and film, TFE produces a smooth, rigid, and transparent film throughout the entire range of flow rates. It has been shown in chapter 4, figure 4.8, that addition of halogenated hydrocarbon to ethane or ethylene causes a rapid increase of the rate of plasma deposition. As an example, addition of 10 cm³/min Freon 12 to a flow of 50 cm³/min of methane increased the polymer deposition rate fiftyfold over that for methane alone. The resulting polymer was found to contain only very small amounts of chlorine. Addition of halogens has also been found [35,36] to produce an increase in the rate of polymerization of styrene and methylmethacrylate. The

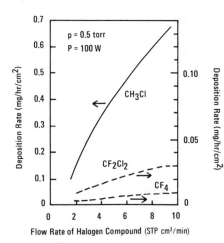

Figure 5.16. Effects of addition of CF_2Cl_2, CH_3Cl, and CF_4 on the deposition rate of methane. *Source:* Adapted from H. Kobayashi, M. Shen, and A. T. Bell, *J. Macromol. Sci., Chem.*, *A8*(8), 1345 (1974). Reprinted by courtesy of Marcel Dekker, Inc., N.Y.

effects of adding CF_2Cl_2, CH_3Cl, and CF_4 on the polymer deposition of methane are shown in figure 5.16.

While both CF_2Cl_2 and CH_3Cl affect the deposition rate very significantly, albeit the latter to a smaller extent, addition of CF_4 has only a small effect. When these three monomers were subjected to an r.f. discharge under the same conditions and their deposition rates plotted against their monomer flow rates (figure 5.17) [34], only CH_3Cl was found to polymerize at an appreciable rate. The authors [34] suggest from these results that both CF_2Cl_2 and CF_4 behave as gas-phase catalysts in methane deposition, while CH_3Cl acts both as a catalyst and a monomer.

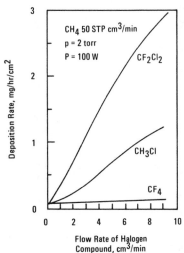

Figure 5.17. Effect of flow rate on deposition rates of CF_2Cl_2, CH_3Cl, and CF_4. *Source:* Adapted from H. Kobayashi, M. Shen, and A. T. Bell, *J. Macromol. Sci., Chem.*, *A8*(8), 1345 (1974). Reprinted by courtesy of Marcel Dekker, Inc., N.Y.

Chapter 4 discussed how the higher rate of deposition of halogenated hydrocarbons (as compared with that of the unhalogenated homologs) appears to be related to the presence of halogen atoms in the gas phase.

5. Effect of Carrier Gases on Deposition

Yasuda and Lamaze [37] reported on the effect of N_2, H_2, He, and Ar on the polymerization of styrene in an electrodeless glow discharge. Some of their results are summarized in table 5.6. Evidently, Ar and N_2 have the largest effect on styrene plasma deposition under the cited conditions. Considering that monomer alone would initiate polymerization in glow discharges and that gas plasma (H_2, N_2, He, Ar) would initiate polymerization of styrene, the data can be represented by the expression [37]:

$$\text{rate} = a[p_m]^2[1 + b(p_x)] \qquad \text{(Eq. 5.28)}$$

Table 5.6. Effect of gas on styrene deposition on glass slide (25×75mm) at 30 watts, 20 min exposure

Composition of plasma	Partial pressure (μ Hg)		Deposited weight (mg)
	Gas	Monomer	
Styrene	0	20	0.20
Styrene-helium	11	20	0.35
	22	20	0.50
	30	20	0.80
	52	20	0.70
	70	20	0.70
	80	20	1.00
	112	20	1.20
Styrene-nitrogen	20	20	0.6
	40	20	0.90
	60	20	1.30
	80	20	1.65
	98	20	1.80
Styrene-hydrogen	20	20	0.30
	30	20	0.40
	45	20	0.50
	60	20	0.55
	70	20	0.60
	80	20	0.60
	100	20	0.45
	110	20	0.45
Styrene-argon	40	20	0.80
	60	20	1.25

Source: H. Yasuda and C. E. Lamaze, *J. Appl. Polym. Sci.*, *15*, 227 (1971). Copyright © 1971, John Wiley & Sons, Inc. Reprinted by permission of John Wiley & Sons, Inc.

where a and b are constants, and p_m and p_x are monomer pressure and gas pressure, respectively. From equation 5.28:

$$R/R_0 = 1 + b(p_x) \qquad \text{(Eq. 5.29)}$$

where R_0 is the deposition rate of pure monomer at given pressure, R is the deposition rate with the mixture of monomer (at the same pressure) and a gas, and p_x is the gas pressure. This expression agrees with the experimental data shown in figure 5.18. The authors [37] observed that the value of b in equation 5.29 is proportional to the relative efficiency of energy transfer of the plasma gas to initiate polymerization of styrene, that is, $k_2 k_3/k_1$. The magnitude of b appeared to be proportional to the cube root of the molecular weight of the gas (table 5.7). (Reaction conditions: $p_m = 20 \mu$ Hg, discharge power = 30 watts, $R/R_0 = 1 + b[p_x]$, where R = polymerization rate with partial pressure, p_x, of gas, and R_0 = polymerization rate of pure monomer at a given pressure).

Abnormally high deposition rates were also observed in a r.f. glow discharge of acetylene and high levels of N_2 flow [38]. This effect was attributed to the change of distribution of polymer deposition due to reduced diffusional transport distance at the higher discharge pressure observed at high N_2/C_2H_2

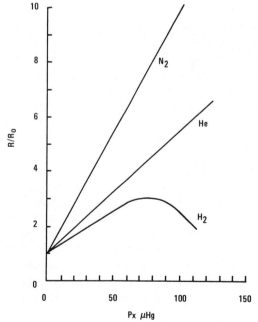

Figure 5.18. Effect of polymer deposition rate on partial gas pressure. The relative deposition rate R/R_0 is according to equation 5.29. *Source:* Adapted from H. Yasuda and C. E. Lamaze, *J. Appl. Polym. Sci., 15,* 227 (1971). Copyright © 1971, John Wiley & Sons, Inc. Reprinted by permission of John Wiley & Sons, Inc.

Table 5.7. Constant *b* for various gases

Gas	$b \times 10^2$, $(\mu\,Hg)^{-1}$
H_2	3.3
He	(4.5)
Ar	(8.5)
N_2	9.0

Source: H. Yasuda and C. E. Lamaze, *J. Appl. Polym. Sci., 15,* 227 (1971). Copyright © 1971, John Wiley & Sons, Inc. Reprinted by permission of John Wiley & Sons, Inc.

ratios. This assumption cannot be tested by increasing the monomer pressure in the absence of N_2, since the system pressure of acetylene in glow discharge cannot be readily manipulated. Acetylene polymerizes at such a rapid rate with evolution of only small amounts of hydrogen that the system pressure remains very low.

Yasuda and Hirotsu [38] showed that at optimum monomer inlet distance, the presence of H_2 as well as Ar approximately doubled the polymer deposition rate of acetylene, while N_2 increased this rate from ~400 Å/min to ~1500 Å/min. The molar ratios of gas to acetylene (Ac) were: H_2/Ac = 1.3, Ar/Ac = 0.8, and N_2/Ac = 1.0.

N_2 and H_2 are known to copolymerize with acetylene. Therefore, not all N_2 and H_2, added in those experiments to acetylene, are considered carrier gases. It is assumed that the high increase of deposition rate of an acetylene-N_2 mixture is partly due to copolymerization of two gases of similar weight (C_2H_2 = 26; N_2 = 28). The effect of carrier gases on ethylene deposition is less pronounced than that on acetylene. Thus, at an optimum monomer inlet distance, the deposition rates in Å/min for ethylene (Et), H_2/Et, Ar/Et, and N_2/Et were approximately 320, 360, 410, and 600 Å/min, respectively. An important aspect of the effect of carrier gas is that it causes a significant narrowing of the distribution of polymer deposition within the reactor (figure 5.19).

The effect of carrier gas added to ethylene, $F = 2.5$ cm³/min, is illustrated in figure 5.20. It can be seen that addition of N_2 on Ar causes a narrowing of distribution. The effects are much less pronounced than those observed with acetylene, however; that is, the distribution is much wider. As would be expected from the reactivities of these two monomers, the maximum in deposition rate for ethylene occurs farther away from the monomer inlet tube than that shown with acetylene. It is also noted that the location of the maximum appears to be independent of the type of carrier gas employed. Furthermore, data provided by workers using differently shaped reactors [39,40] indicate

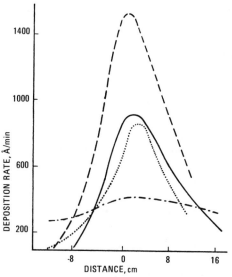

Figure 5.19. Distribution of polymer deposition in plasma polymerization of acetylene with addition of a gas. *Source:* Adapted from H. Yasuda and T. Hirotsu, *J. Polym. Sci., Polym. Chem. Ed., 16,* 229 (1978). Copyright © 1978, John Wiley & Sons, Inc. Reprinted by permission of John Wiley & Sons, Inc.

——., acetylene (Ac), $p_g = 8$ μm Hg
....., H_2/Ac, $p_g = 70$ μm Hg
——, Ar/Ac, $p_g = 8.8$ μm Hg
-----, N_2/Ac, $p_g = 45$ μm Hg

that the position of the maximum in deposition rate is dependent, in addition to the rate of flow, on the geometry of the reaction vessel.

The aforementioned trends in the effect of carrier gases suggest examination of two factors that appear to be most important in these processes. These are the rate by which monomer molecules are converted to the monomer plasma and the rate by which polymers are formed in the plasma. It may be assumed that the energy required to convert monomer molecules to the monomer plasma is carried chiefly by nonpolymerizing species such as the small amounts of hydrogen generated in the process of plasma polymerization of hydrocarbons or by carrier gas.

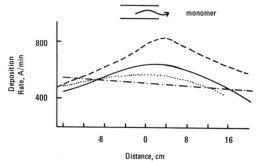

Figure 5.20. Distribution of polymer deposition in plasma polymerization of ethylene with addition of gas. *Source:* Adapted from H. Yasuda and T. Hirotsu, *J. Polym. Sci., Polym. Chem. Ed., 16,* 229 (1978). Copyright © 1978, John Wiley & Sons, Inc. Reprinted by permission of John Wiley & Sons, Inc.

——., Et, $p_g = 44$ μm Hg
....., H_2/Et, $p_g = 87$ μm Hg
——, Ar/Et, $p_g = 110$ μm Hg
-----, N_2/Et, $p_g = 96$ μm Hg

Yasuda and Hirotsu [38] considered the following factors important in determining the distribution of polymer deposition:

Monomer reactivity in the plasma state.

The frequency of reacting collisions, which depends on pressure and molecular weight.

The diffusional displacement distance, which is related to the numbers and the velocity of the reactive species.

The monomer flow rate.

The presence and type of nonreacting gas molecules and the frequency of nonreacting collisions.

Ratio of surface to volume of a reactor.

Ethylene, which is less reactive than acetylene, travels a longer distance randomly throughout the reaction tube before being involved in deposition processes. Acetylene, on the other hand, tends to deposit quickly in the vicinity of the monomer inlet.

The diffusional displacement distance of a gas is given by Einstein as:

$$\bar{X}^2 = 2Dt \qquad \text{(Eq. 5.30)}$$

where $\bar{X}$ is the one-dimensional displacement executed by a particle during time t in a medium with a diffusion coefficient D. At a low pressure, p, the diffusion coefficient can be calculated [41] from the diffusion coefficient, D_0, at the standard state ($p_0 = 760$ mm Hg and $T_0 = 273°$K) by

$$D = D_0 \, (T/T_0)^n \, (p_0/p) \qquad \text{(Eq. 5.31)}$$

where n lies between 1.75 and 2.0. Consequently, the diffusional displacement distance is shorter for higher-molecular-weight molecules of low D_0 value and for the system in higher pressure. Therefore, the greater the reactivity of monomer plasma and/or the higher the system pressure, the narrower would be the distribution of polymer deposition. The effect of carrier gas to narrow the distribution appears to be related to the mass of gas molecules. The effect is on the order of $N_2 > Ar > H_2$.

While the aforementioned data point to the molecular size or mass of the carrier gas as a factor that determines the deposition rate, it is also known [42] that the large molecule Freon 12, CCl_2F_2, in concentrations of 10%, increased the deposition rate of ethylene in r.f. discharge by a factor of 10. This effect attributed to a catalytic action rather than to molecular mass. On the other hand, it has been shown by Yasuda and Hirotsu [38] that the enhancement effect of deposition as observed on Freon 12 is due to copolymerization with ethylene.

Since monomer is supplied to the reactor in the vapor phase, the mass transport of monomer should remain an important factor regardless of reactor systems. On the other hand, the mode of energy input needed to maintain a system in the plasma state, as well as its relative position to the location of polymer deposition, is considered [38] to be another important factor that modulates the deposition distribution. It should be noted that, under conditions of polymer distribution on an electrode surface, the effect of energy transfer rate is absent, and therefore the main effect appears to be the mass transfer of monomer.

The importance of the effect of carrier gas on the distribution of deposition rates lies in the fact that maximum deposition can be localized for a given specific objective. The deposition rate that is observed at a strategically placed position, such as on the electrode surface, in the space between electrodes, or on the wall of the reactor, can be altered to suit a given target.

6. Effect of Discharge Power W, W/F, and W/FM

It has been shown in section 3 of this chapter that in the r.f. glow-discharge polymerization of styrene [25], the rate of free-radical production, r, is proportional to $W^{1.7}$, which suggests that radicals are being generated by both first- and second-order processes in power, W, that is,

$$r = k''W^2 + k'W \qquad \text{(Eq. 5.32)}$$

or

$$r/W = k''W + k' \qquad \text{(Eq. 5.33)}$$

Thus, a plot of r/W against W yielded the values $k'' = 5.17$ and $k' = 3.30$, and the rate of polymer deposition, R, was given as [25]:

$$\text{Rate} = \frac{p(k''W^2 + k'W)}{p + A + [A^2 + B(k''W^2 + k'W)]^{\frac{1}{2}}} \qquad \text{(Eq. 5.34)}$$

where p = pressure in the system, and A and B are ratios of the specific rate constants for the reactions $R_{\dot{n}} + M$, $R_{\dot{n}} + R_{\dot{m}}$ and $R_{\dot{n}} \rightarrow$ (trapped), that is, $A = k_3/2kk$; $B = k_2/k^2k_1$. Their values were [25]: $A = 0.277$, B 0.011. Under the same plasma conditions, the A and B values for α-methylstyrene and allylbenzene were [43]:

α-methylstyrene: $A = 0.20$, $B = 0$
allylbenzene: $A = 0.55$, $B = 0$

In these two latter cases, a value of zero for constant B indicates that little or no recombination reactions in these two monomer plasma environments occurred. Denaro and colleagues [43] obtained the following values for the rate of radical production, r, and the rate of deposition, R:

α-methylstyrene $r = 12.3W$
$R = 9.32W$
allylbenzene $r = 12.0W$
$R = 6.38W$

Subsequently, the authors reported [44] the values for allyl alcohol as $r = 5.82W^{0.57}$ and $R = 1.92W^{0.57}$, indicating that any relationship between the rate of radical formation and discharge power is extremely complicated. These investigators interpreted their experimental results in terms of a mechanism involving the competitive reactions of free radicals formed on the electrodes by electron bombardment. They considered the propagation step to involve the reaction of radicals with adsorbed monomer and with monomer diffusing to the electrodes from the gas phase. The power was fed at 2 MHz to parallel electrodes; the experiments were conducted in a pressure range between 0.2 and 2.0 torr.

K. C. Brown [45], studying a number of vinyl halides under similar experimental conditions, suggested that electrons or ions from the discharge generate radicals and ions from monomer adsorbed on the electrodes. Figure 5.21 shows the variation of deposition rate with power density. The flow rates used to determine the power curves were those that gave the maximum deposition rate and were similar for all monomers investigated.

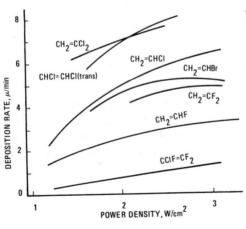

Figure 5.21. Deposition rate versus power density for vinyl halides, pressure 1.9 torr, temperature 15°C, frequency 3.14 MHz. *Source:* Adapted from K. C. Brown, *Europ. Polym. J.*, *8*, 117 (1972). Reprinted with permission from *Europ. Polym. J.* Copyright 1972, Pergamon Press, Ltd.

The dihaloethylene polymerized at a higher rate than the corresponding monohalides; the chlorides and bromides polymerized more quickly than the fluorides. There were no significant differences between the deposition rates of the two dichloroethylenes. The tetrahalide, chlorotrifluoroethylene, polymerized very slowly, as tetrafluoroethylene has been found to do.

More recently, Bell and his coworkers [46] polymerized methane, ethane, propane, and n-butane in a r.f. plasma. They observed that, in addition to transparent films, powders may be formed at low pressures and low monomer flow rates; unstable discharges were obtained at high pressures and low flow rates. With increasing power input, the unstable regions are decreased, while the powdery regions are increased. The rates of polymer deposition were found to depend on pressure, flow rate, and power. Figure 5.22 shows the dependence of deposition rate on power input.

Because of increasing power input, a larger number of active species would be expected to be present in the plasma and would therefore increase rates of deposition. The data are in agreement with this assumption, the exception being methane. Here, a difference in mechanism is suggested for methane [46]. Somewhat earlier it was proposed [47] that acetylene is formed as a reaction intermediate in the plasma polymerization of ethylene:

$$e + C_2H_4 = C_2H_2 + 2 H \text{ (or } H_2) + e \qquad \text{(Eq. 5.35)}$$

Under comparable conditions, ethylene polymerizes in a plasma at a slower rate than acetylene by about an order of magnitude. The plasma polymerization of ethane is in turn slower than that of ethylene. It is assumed that an extra

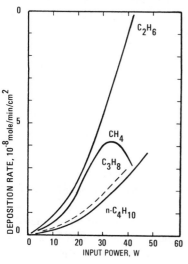

Figure 5.22. Deposition rate versus power input for methane, ethane, propane, and n-butane. Flow rate 10 cc/min (S.T.P.), pressure 2.5 torr. *Source:* Adapted from H. Hiratsuka, G. A. Kovali, M. Shen, and A. T. Bell, *J. Appl. Polym. Sci.*, 22, 917 (1978). Copyright © 1978, John Wiley & Sons, Inc. Reprinted by permission of John Wiley & Sons, Inc.

step is needed for the initial decomposition of ethane by electron impact to form ethylene first,

$$e + C_2H_6 = C_2H_4 + H_2 + e \qquad \text{(Eq. 5.36)}$$

followed by reaction as shown in equation 5.35.

The investigators suggested that a similar mechanism may also be operative for propane and n-butane, but in the case of methane, the absence of the C-C bond in the monomer excludes this type of mechanism. They cite photochemical data showing that the initial decomposition to diradicals is required for the formation of ethylene from methane and suggest that this difference in mechanism may be responsible for the observed difference in the kinetics of methane polymerization in a glow discharge.

Thompson and Mayhan [48], who in 1972 postulated a cationic polymerization scheme for plasma polymerization, showed that the mass of polymer formed in an inductively coupled plasma increases with power at any point in the reactor and that the position of the maximum rate of formation, that is, at $dp/dx = 0$, shifts inward toward the r.f. field as the power level increases. Plasma was sustained above 8 watt power throughout the reactor length. (See figure 4.7.)

More recently, Yasuda and Hsu [49] reported that in the plasma polymerization of ethylene and tetrafluoroethylene, glow characteristics play an important role and that these depend on the combined quantity W/F_m, where W is the discharge power and F_m is the monomer flow rate. At higher flow rates, higher wattages are needed to sustain a "full glow." Above a certain value of W/F_m decomposition becomes the predominant reaction, and the polymer deposition rate decreases with increasing discharge power. ESCA data indicate that when polymerized in an incomplete glow region, that is, low W/F_m, the plasma polymer of tetrafluoroethylene is a hybrid of linear and plasma polymers. Under conditions of high W/F_m, that is, "full glow," polymers are alike, regardless of the extent of monomer decomposition. They contain carbons with different numbers of F, such as

$$CF_3, \ -CF_2-, \ \diagdown CF \diagdown C \diagup ,$$

as well as carbons bonded to other, more electronegative substituents. The effect of deposition rate on discharge power is illustrated in figure 5.23 for three flow rates.

Soon afterward, Yasuda and Hirotsu [50] extended this concept to include the molecular weight, M, of the monomer. These authors observed that, at a fixed flow rate, the discharge power has a relatively small effect on the

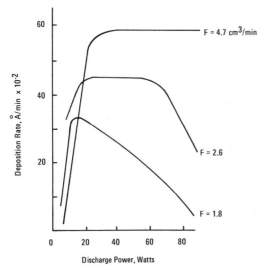

Figure 5.23. Polymer deposition rate versus discharge power at various flow rates for tetrafluoroethylene. *Source:* Adapted from H. Yasuda and T. S. Hsu, *J. Polym. Sci., Polym. Chem. Ed., 16,* 415 (1978). Copyright © 1978, John Wiley & Sons, Inc. Reprinted by permission of John Wiley & Sons, Inc.

distribution pattern of polymer deposition as long as the values of W/FM are maintained above a critical level to maintain full glow in the reactor.

Plasma polymerization of two monomers of different molecular weights can be compared by selecting conditions that yield similar W/FM values. Data of these authors appear to confirm the following aspects of plasma polymerization [50].

Polymer deposition rates as a function of flow rate at fixed power or as a function of power at fixed flow rate may only reflect the change of deposition pattern and may not correspond to the change of polymerization rate.

The deposition rate is a complex function of polymer conversion yield, the intrinsic polymerization rate of a monomer, and molecular weight of the polymer-forming unit, which may not necessarily be identical with the monomer. Under conditions of nearly 100% conversion and controlled discharge power to maintain full glow, the deposition rate is directly proportional to the monomer molecular weight.

The value of W/FM seems to provide a first-order approach to obtain a fair comparison of different monomers, especially of monomer homologs. In the power range close to the critical value of W/FM, the effect of system pressure tends to become noticeable.

The most uniform distribution of polymer deposition is obtained at relatively low flow rates and a discharge power above the critical value of W/FM [51]. Therefore, these conditions will be most suitable when polymer deposition rates of various monomers are compared without carrying on a distribution study.

7. Effect of Electrode Temperature on Deposition

Williams and Hayes [22] suggested in 1966 that all "difficult" monomers would probably polymerize quite readily on electrodes maintained at a suitable temperature. These workers cite ethylene as an example of a monomer that polymerizes slowly at room temperature but would probably react quite rapidly on an electrode maintained at a temperature of about $-200°C$. Westwood [23] demonstrated that, in a r.f. discharge, the deposition rates for vinyl chloride, vinyl acetate, and methyl methacrylate decrease with rising temperatures while keeping other parameters, such as system pressure, flow rate, and current density, constant (figure 5.24).

The relation between gas pressure and the surface concentration of adsorbed molecules is determined by the adsorption isotherm at the electrode temperature. The parameter largely controlling the degree of adsorption at a given temperature is the ratio of gas pressure, p, to the saturated vapor pressure, p_0, at that temperature. It seems, therefore, informative to plot the rate of polymer deposition as a function of relative pressure, p/p_0. For vinyl acetate, the plot is shown in figure 5.25 [23]. These experiments were carried out at a constant gas pressure, p, with an effectively varying p_0. By contrast, in the case of an adsorption isotherm, p is varied while p_0 is kept constant. The dependence of polymer deposition rate on the substrate temperature is

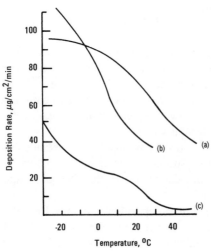

Figure 5.24. Effect of temperature on polymer deposition: (*a*) vinyl chloride, 0.99 torr, 1.53 mA cm^{-2}, 163 V^{-1}. (*b*) methylmethacrylate, 1.96 torr, 2.43 mA cm^{-2}, 205 V cm^{-1}; (*c*) vinyl acetate, 0.99 torr, 2.43 mA cm^{-2}, 113 V cm^{-1}. *Source:* Adapted from A. R. Westwood, *Europ. Polym. J.*, *7*, 363 (1971). Reprinted with permission from *Europ. Polym. J.* Copyright 1971, Pergamon Press, Ltd.

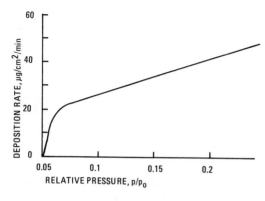

Figure 5.25. Effect of relative pressure on deposition rate for vinyl acetate. *Source:* Adapted from A. R. Westwood, *Europ. Polym. J.*, *7*, 363 (1971). Reprinted with permission from *Europ. Polym. J.* Copyright 1971, Pergamon Press, Ltd.

shown for 2-vinylpyridine in figure 5.26 at a constant pressure of 100 millitorr and constant current density of 0.1 mA/cm² [26].

Halovka [26] determined that the overall rate expression for the formation of polymer, R, from 2-vinylpyridine on the electrode surfaces is given by

$$R = \frac{2270j(P_{2\text{-VP}} + 94)}{T} \qquad \text{(Eq. 5.37)}$$

where R is in units of angstrom/min, the current density, j, in mA/cm², the monomer pressure, $P_{2\text{-VP}}$, for 80 to 180 millitorr, and the temperature, T, in °C. Thus, the rate of deposition of plasma PVP is inversely proportional to the temperature, suggesting that adsorption processes are important in the reac-

Figure 5.26. Effect of substrate temperature on deposition rate for 2-vinylpyridine. *Source:* Adapted from J. M. Halovka, Sandia Labs., Res. Rept. No. SLA 74-0181 (1974).

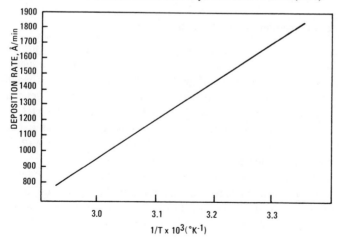

Table 5.8. Effect of electrode temperature on polymer deposition for methyl methacrylate

Electrode temp. (°C)	Monomer pressure (torr)	Flow rate (cm³/min)	Deposition rate (microns/min)
−6	1.9	7.0	1.0
15	1.9	7.0	0.9
25	1.9	6.8	0.7
40	1.9	7.0	0.45

Source: K. C. Brown, *Europ. Polym. J., 8,* 117 (1972). Reprinted with permission from *European Polymer Journal.* Copyright © 1972, Pergamon Press, Ltd.

tion mechanism. It is noted that the overall rate expression was unaffected by the partial pressure of carrier gas argon.

Plasma deposition of methyl methacrylate at 3.14 MHz r.f., 1.9 torr, and a flow rate of 6–7 cm³/min was also shown [45] to decrease with increasing temperature (table 5.8).

The films formed at low temperature were found to be softer and stickier than corresponding films at higher temperatures. When the films formed at various temperatures were heated to 100°C for several hours, they hardened and lost a considerable amount of weight in the process. These losses appear to be due to monomer entrapped in the deposition process. It is interesting to note, however, that the effect of heating the films to 100°C was also dependent on the monomer pressure in the glow discharge. Thus, at 3.5 torr, the gain of deposition rate at −10°C proved to be no real gain, since deposition, corrected after heating at 100°C, was approximately the same as that measured at 40°C electrode temperature. As the monomer pressure was lowered, the "true" gain in deposition increased.

8. Effect of Frequency on Deposition

A comparative study was recently made [52] to compare the effect of plasma polymerization of benzene in microwave and radio-frequency glow discharge. The authors used identical reactor design as well as reaction parameters of pressure, flow rate, and power. The microwave generator and radio-frequency generator operated at 2540 MHz and 13.56 MHz, respectively. The power/pressure ratio, W/P, was considered to be an indication of the energy supplied to the plasma; the greater the value of the W/P ratio, the more energetic the plasma.

The authors reported that films obtained with values of W/P larger than 200 are insoluble in most solvents, indicating a high degree of crosslinking. The compounds contained in soluble films covered a broad range of molecular weights. It is shown from the gel permeation chromatography (GPC) spectra in figure 5.27 that the soluble compounds of high-molecular-weight materials with sizes related to those of terphenyl and substituted benzenes are more important in the case of microwave discharges. Furthermore, infrared spectra of the films showed that the aromatic character decreases as the discharges become more energetic; it is lower in microwave than in r.f. discharges. The aromatic character of polymers obtained in low-energy r.f. discharges is reported to be close to that of conventional polystyrene.

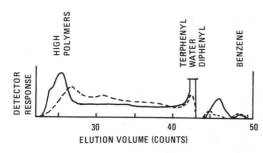

ELUTION VOLUME (COUNTS)

Figure 5.27. GPC spectra of plasma-produced polymers. ————, microwave. _ _ _ _, radio frequency. Concentration 0.2% THF. *Source:* Adapted from M. Duval and A. Théorêt, *J. Electrochem. Soc., Solid State Sci. and Technol., 122* (4), 581 (1975). Reprinted by permission of the publisher, The Electrochemical Society, Inc.

This finding has been confirmed by other investigators [53,54,55]. Polymers with large differences of aromatic character, however, may also be obtained in both microwave and r.f. discharges by using different W/P ratios [52]. In addition the IR spectra indicated that the aromatic structures contain pendant groups of aliphatic or alicyclic nature. In low-energy discharges, these are relatively strong bonds at 3300 cm^{-1}, indicative of C≡C triple bonds.

When the main aromatic bonds are missing or are weak, as observed mainly in high-energy microwave discharges, carbon-carbon double-bond structure becomes more prevalent. Plasma emission spectra observed through a quartz window are represented in figure 5.28. They contain the prominent features of all other spectra reported. Monoatomic, diatomic, and polyatomic radicals or ions, as well as excited molecules, are responsible for these emission spectra. They are gas-phase unstable species that do not relate to the more stable radicals that may be trapped in polymer films.

Those species which have been identified include bands at:

4350, 4860, 6580 Å	Main Balmer lines of H atom
3900, 4310 Å	CH radical transitions
4310, 5160, 5630 Å	Swan bands due to transition of the C_2 radical
2400–3300 Å region	Transition of benzene molecule
4673–6500 Å region	Due to emission of C_4H_2 or C_2H radical
4456–5900 Å and	Assigned to emission of $C_6H_5CH_2$ and C_6H_4
3400–4400 Å	radicals

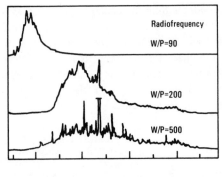

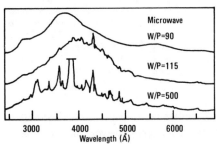

Figure 5.28. Emission spectra of radio-frequency and microwave discharges for several values of W/P. *Source:* Adapted from M. Duval and A. Théorêt, *J. Electrochem. Soc., Solid State Sci. and Technol., 122,* (4), 581 (1975). Reprinted by permission of the publisher, The Electrochemical Society, Inc.

Drawing on their data, the investigators [52] postulated the following:

In microwave discharges the main emission comes from CH radicals and, to a lesser extent, from C_2 and C_2H radicals.

In high-energy r.f. discharges the main reactive species are H, C_2H and, to a smaller extent, $C_6H_5CH_2$ and C_6H_4 radicals.

In low-energy discharges, ring fragmentation is replaced by ring excitation, particularly in the case of r.f. discharges.

In intermediate-energy discharges, there is a continuous variation, as a function of W/P, of the amount of CH and H radicals.

9. Effect of Reactor Design

When a steady state of monomer flow is established at a given system pressure, p_0, and a glow discharge is initiated, the system pressure in the glow discharges, p_g, will reach a constant value within a short period of time. The value of p_g is generally different from p_0 and is largely dependent upon the nature of monomer, flow rate, and pumping rate. For readily polymerizing monomers, such as acetylene, ethylene, and acrylonitrile, the pressure in the glow zone, p_g, is dependent on the pressure of the gaseous products, such as hydrogen. If the reactor design is such that the monomer enters a horizontal tube at the side, the gas phase at the end of the reactor may consist of hydrogen only. This condition is comparable to gas entering a flame in a gas burner, in which combustion rate and flow rate of gas establish a steady-state flame. In the above example, monomers will interact immediately with the active species generated by the r.f. coil and are then transported by diffusion.

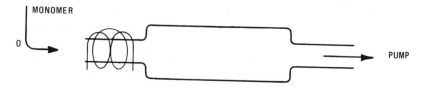

If the polymer-forming reactivity and the feed rate of the monomer are both high, polymer tends to deposit near the monomer inlet; that is, the monomer requires low residence times in the plasma to form polymer. It is noted that the monomer forms reactive gases in the plasma state that yield (nongaseous) polymer plus hydrogen and/or other products of fragmentation.

As an example, acetylene polymerizes in plasma at a high rate under evolution of small amounts of hydrogen. As a result, after a steady-state plasma flow is established, the system pressure in glow discharge, p_g, approaches zero and can be maintained at low pressure for some time. Generally, active species, such as electrons, H atoms, and excited hydrogen molecules, are chiefly generated under the r.f. coil and are transported by diffusion through the entire reactor volume, which is assumed to be under full glow. As monomer enters the glow discharge, the monomer becomes a polymer-forming, energy-consuming species, M^*, that is, the monomer plasma. Its Brownian motion is high at the time of excitation and is inversely proportional to the monomer's molecular weight.

At very low flow rates the residence time of the gas in the discharge is sufficiently high to establish a steady state of concentration of reactive species, such as free radicals and ions, which are independent of the gas flow rate. These reactive species subsequently initiate both homogeneous and heterogeneous polymerization via mechanisms discussed in chapter 4. For sufficiently long residence times, most or all of the monomer introduced will be converted to polymer. As the flow rate is increased, the polymerization rate increases initially because of the greater supply of monomer. Finally, a point is reached at which the flow rate begins to perturb and reduce the concentration of the reactive plasma species (free radicals, ions). This event, together with the decreased residence time, results in a situation in which only a fraction of the monomer feed can be converted.

It should be noted that polymer deposition rate is not identical with polymerization rate. The discussion in this chapter is confined to deposition rates. The rate of deposition at a given site in the reactor depends on the distribution of polymer deposition in a given system. In return, it is influenced by such factors as the diffusional transport of active species generated in the region of energy input; that is, under the r.f. coil, the diffusional transport of polymer-forming species, and the flow of product gas and monomer.

Yasuda and Hirotsu [56] have conducted a most revealing study on the distribution of polymer deposition by altering the position of monomer inlet as well as the direction of monomer flow in their reactor, using ethylene as monomer. The flow patterns are schematically shown in figure 5.29.

In this reactor design, the diameter of the tube portion that is surrounded by the r.f. coil was made smaller than that of the main part of the reaction tube. In this manner the amount of polymer deposition under the coil was reduced so that a large-volume tail-flame portion of glow discharge could accommodate plasma polymerization. The intent was to decouple the initial impact of electron bombardment and thus reduce many secondary effects that may impair the demonstrative aspect of the study. The distribution of polymer deposition is expressed by the distance from an arbitrary zero point to the right of the center of the reactor, as indicated. Aluminum foils, 1 × 2 cm, at 4 cm intervals, were placed on a glass plate 2.5 × 38 cm and their weight increases determined as measurement of the deposition rate.

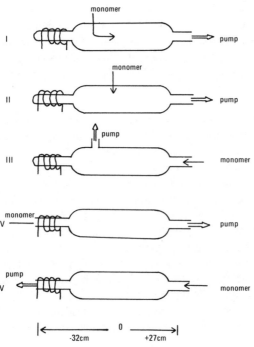

Figure 5.29. Schematic representation of flow patterns. *Source:* Adapted from H. Yasuda and T. Hirotsu, *J. Polym. Sci., Polym. Chem. Ed., 16,* 313 (1978). Copyright © 1978, John Wiley & Sons, Inc. Reprinted by permission of John Wiley & Sons, Inc.

The results of these experiments show no significant differences in the deposition pattern in cases I and II, where the ethylene was injected into the direction of the flow and perpendicular to it, respectively. When the direction of flow was reversed (cases III and V), however, a drastic change in the distribution pattern was observed, as shown in figure 5.30. When the flow direction is reversed, the glow does not extend through the entire length of the reaction tube; therefore no polymer deposition occurs beyond +8 cm. The glow was observed to penetrate into the outlet tube in case III, however.

It is interesting to note that the authors [56] found no marked differences in the polymer deposition pattern when comparing cases III and V. But when the flow passes through the r.f. coil, a heavy deposition occurs in the vicinity of the r.f. coil, thus reducing the extent of deposition elsewhere. In these two cases, the polymer deposition depends largely on the direction of the flow. When the reactor is located on the downstream side of the r.f. coil, the glow is found to extend through the entire length of the tube.

With a.c. discharges on electrodes, best results have been obtained [25] at higher frequencies. As the frequency of the discharge is increased, more energy is transferred to the gas and less to the electrodes, the maximum energy being transferred to the gas when the angular frequency, ω, of the supply is equal to the collision frequency, ν, between electrons and the gas.

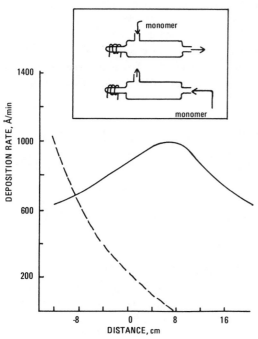

Figure 5.30. Effect of monomer flow reversal on polymer deposition. *Source:* Adapted from H. Yasuda and T. Hirotsu, *J. Polym. Sci., Polym. Chem. Ed., 16*, 313 (1978). Copyright © 1978, John Wiley & Sons, Inc. Reprinted by permission of John Wiley & Sons, Inc.

At a pressure of 1 torr, $v \simeq 10^9$ sec^{-1}, corresponds to a discharge frequency of about 160 MHz.

10. Effect of a Magnetic Field

Plasma deposition in a system containing two electrodes inside the reactor has generally been measured by weighing the polymer deposit accumulated on the electrodes; it has been shown in section 3B of this chapter that films deposited at a lower pressure are of higher quality than those formed at higher pressures, although they were deposited at a slower rate.

Recently, Yasuda and his coworkers [57] conducted an interesting experiment in a low-pressure plasma by measuring the polymer deposition midway between electrodes. Using a magnetic field they confined the plasma in the interelectrode gap near the surface of the electrode. The relative orientation of substrate and electrodes in the resistively coupled glow discharge is shown schematically in figure 5.31.

An iron frame was attached to, but isolated from, the back of each aluminum electrode. A small disk, S, was fixed to the back of the electrode. Four permanent magnets were positioned with their north poles at the edges of the iron frame and south poles on the iron disk so that the north-south vectors were oriented at 90° intervals. In this way, the investigators arranged a donut-shaped magnetic field projecting into the electrode gap to prevent the escape of electrons to the outside of the interelectrode region. In this arrangement, the electrons are also forced into a convoluted path from one electrode to the other, thus enhancing the probability of electron-molecule collisions and concentrating the plasma between electrodes. Deposition rates were measured at 60 Hz (r.f.), 10 kHz (a.f.), and 13.5 MHz (r.f.) at various locations on the substrate and electrode.

Figure 5.31. Schematic representation of the electrodes and substrate assembly: (*left*) side view; (*right*) front view. Magnets are placed with their north poles on the frame, *N,* and south poles on the circle, *S.* The assembly is placed in a bell jar. Dimensions are in cm. *Source:* Adapted from N. Morosoff, W. Newton, and H. Yasuda, *J. Vac. Sci., Technol., 15*(6) (Nov./Dec. 1978).

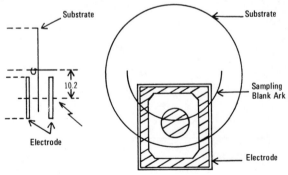

The investigators reported [57] that such magnetic enhancement results in a ring of intense glow near the electrode with its center coinciding with an axis passing through the electrode centers. At 60 Hz and 10 kHz the ring is narrow and well defined, but at 13.56 MHz it is diffuse. It was observed that in such an arrangement the magnetic enhancement permitted a greater latitude of power applied to the glow discharge without the typical arcing that occurs in

Figure 5.32. Deposition rate profile on the electrode and substrate for plasma, 13.56 MHz, ethylene flow 6.15 cm³/min, pressure, 30 m torr: (*top*) with magnetic enhancement; (*bottom*) without magnets. *Source:* Adapted from N. Morosoff, W. Newton, and H. Yasuda, *J. Vac. Sci. Technol.*, *15*(6) (Nov./Dec. 1978).

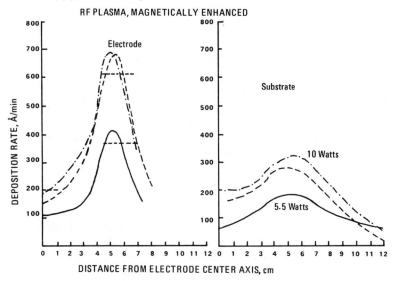

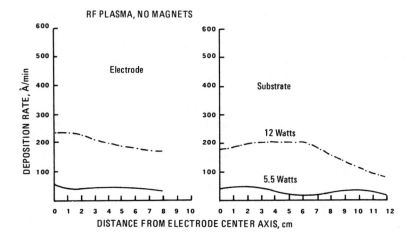

the absence of the magnetic field. The results are illustrated in figure 5.32 for ethylene.

The data of figure 5.32 indicate clearly that high deposition rates result from such magnetic fields. In the absence of the magnetic field, a high potential drop across the electrodes would be necessary to initiate a glow discharge under these conditions, and such an increase in the current level would result in arcing. Presence of such magnetic enhancement reduces the potential drop required for a given current level while increasing the deposition rate on both the electrode and substrate. Higher pressures were found to reduce the conversion to polymer significantly.

At both a.f. and r.f., the deposition rate was shown to increase with power up to a fairly constant maximum, but it was consistently lower at the audio-frequency level. Without magnetic enhancement, the deposition rates on the substrate were well below those obtained under the effect of a magnetic field at any of the three frequencies employed. The authors note that in spite of the enhanced deposition rate, the adhesion of the plasma polymer to the aluminum specimens is very good from low pressure a.f. and r.f. plasma and superior to that of plasma-polymerized ethylene using inductive coupling. They also point out that a.c. operation under the preferred low-pressure conditions results in excessive heating of the electrodes and should therefore be avoided.

11. Effect of Pulsed Discharge

Little work has been reported on the effect of pulsed power in discharge. In the production of H atoms from molecular hydrogen, Shaw [58] observed no significant difference between continuous and pulsed power at frequencies of 3000–9000 Mc and 5–300 W. This does not seem unexpected, since the lifetime of atoms in plasma is about 0.1–1.0 sec. On the other hand, high concentrations of atoms from N_2 were determined [59] at a pulse rate of 30–60/sec. Apparently, a pulsed type of discharge is capable of generating a higher concentration of nitrogen atoms from pure N_2 than any other system. Yasuda and Hsu [60] were the first to disclose interesting information on the effect of pulsed plasma on the glow-discharge polymerization of organic compounds.

In conventional free-radical polymerization, the rate of radical production is identical to the rate of radical comsumption; therefore, no free radicals are trapped in the polymer formed. In plasma polymerization, on the other hand, the rates of radical production and consumption are not directly related. A different situation exists, however, when the r.f. discharge is supplied in pulsed spurts. One would expect the formation of free radicals to be proportional to the duration of r.f. discharge. While the discharge is off, no radicals are generated, but during this off period, the radicals created during the preceding discharge pulse can interact with each other, with a neutral molecule, or with excited species. Therefore, a pulsed discharge should be expected to alter the balance between generated and consumed free radicals.

Yasuda and Hsu [60] carried out their experiments under steady-state flow conditions. They reasoned that under these conditions, any reaction is associated with a pressure change, δ,

$$\delta = p_g/p_m \qquad \text{(Eq. 5.38)}$$

where p_m is the monomer pressure before initiation of the discharge and p_g is the pressure of the steady-state flow system during the discharge. In a closed system, the value of δ in the plasma polymerization of hydrocarbons would equate to the hydrogen yield. In a flow system, an increase of the δ value may indicate either an enhancement of hydrogen formation or of unreacted monomer.

When acetylene and benzene were exposed to both continuous and pulsed discharges in different experiments in a closed system, there was no identifiable change in the characteristic reactions of polymer formation. Still, the rates of polymer deposition differed significantly as indicated by the half-times of reactions, $t_{\frac{1}{2}}$, in table 5.9. The pulsed r.f. reduces the rate of polymer formation of acetylene to one-fifth and that of benzene to one-eighth.

When the experiments were conducted in a flow system, significant increases in δ values were observed in the pulsed discharge. They are believed to be due to a decrease in the rate of polymer formation. Significant decreases in δ values were obtained, however, for acrylic acid and propionic acid. The deposition rates from both continuous and impulsed discharges are compared for different monomers in table 5.10.

The deposition rates are based on the total time of the pulsed discharge. It can be seen that the rates generally decrease with the pulsed discharge, 100 μsec on, 900 μsec off. On the other hand, considerable increases in the deposition rate have been found for acrylic acid, propionic acid, and tetrafluoroethylene when the pulsed discharge was employed.

Table 5.9. Effect of pulsed r.f. on acetylene and benzene in a closed system

Polymerization parameter	Acetylene		Benzene	
	Continuous	Pulsed	Continuous	Pulsed
$t_{\frac{1}{2}}$, sec	1.0	5.0	2.5	18.5
Hydrogen yield, P_{H_2}/P_0	0.15	0.14	0.13	0.14
Polymer yield, $z = 1 - X$	0.98	0.98	0.99	0.99

Source: H. Yasuda and T. Hsu, *J. Polym. Sci., Polym. Chem. Ed., 15,* 81 (1977). Copyright © 1977, John Wiley & Sons, Inc. Reprinted by permission of John Wiley & Sons, Inc.

Table 5.10. Deposition rates of various monomers in continuous and pulsed r.f. discharges

Monomer	Deposition rate $\times$ 10^8, g/cm^2-min		
	Continuous	Pulsed	Change[a]
C_2H_2	31	24	$-7(-23)$
C_6H_6	110	101	$-9(-8)$
C_6F_6	190	149	$-41(22)$
Styrene	173	145	$-28(-16)$
C_2H_4	42	43	$+1(+2)$
C_2F_4	18	37	$+19(+110)$
Cyclohexane	92	9	$-83(-90)$
Ethylene oxide	15	14	$-1(-7)$
Acrylic acid	28	61	$+33(+120)$
Propionic acid	7	15	$+8(+110)$
Vinyl acetate	31	16	$-15(-48)$
Methyl acrylate	32	33	$+1(+3)$
Hexamethyldisilane	251	65	$-186(-74)$
Tetramethyldisilane	191	102	$-89(-47)$
Hexamethyldisiloxane	233	43	$-190(-82)$
Divinyltetramethyldisiloxane	641	277	$-364(-57)$

[a]Changes are based on values of continuous discharge. Numbers in parentheses are percentages.
Source: H. Yasuda and T. Hsu, *J. Polym. Sci., Polym. Chem. Ed., 15,* 81 (1977). Copyright © 1977, John Wiley & Sons, Inc. Reprinted by permission of John Wiley & Sons, Inc.

Propionic acid polymerization is assumed to proceed chiefly via hydrogen detachment. At the same time, however, carboxylic acid groups are lost. It appears that the pulsed discharge reduces the extent of carboxylic acid detachment, thus increasing the polymer mass deposited per unit time. This interpretation is supported by infrared spectra and by the observed increase in wettability of the pulsed-discharge polymer as compared with the polymer obtained from continuous discharge.

It is interesting to note that unsaturated monomers showed an increase in free-radical concentration during the pulsed discharges and no decreases during the off period. This fact led to the conclusion that a conventional polymerization mechanism is operative during the off period of the pulsed discharge and activated to a plasma-type mechanism during the on period.

The considerable increase in deposition rate of acrylic acid by pulsed discharge may be due to contributions from additional polymerization during the off period, which tends to retain the carboxylic acid group and thus increases the molecular weight of the repeating unit.

References

1. J. S. Townsend. *Electrons in Gases.* Hutchinson Sci. and Techn. Publ.: London (1947).

2. A. V. Phelps. In *Chemical Reactions in Electrical Discharges*, ed. B. D. Blaustein. Advances in Chemistry Ser., *No. 80*. Amer. Chem. Soc.: Washington, D.C. (1969).

3. R. D. Hake, Jr., and A. V. Phelps. *Phys. Rev., 158*, 70 (1967).

4. A. D. McDonald. In *Microwave Breakdown in Gases*. Wiley: New York (1966).

5. P. Felsenthal and J. M. Proud. *Phys. Rev., 139*, A1796 (1965).

6. H. Raether. In *Electron Avalanches and Breakdown in Gases*. Butterworth: Washington, D.C. (1964).

7. W. L. Fite. In *Chemical Reactions in Electrical Discharges*, ed. B. D. Blaustein. Advances in Chemistry Ser., *No. 80*. Amer. Chem. Soc.: Washington, D.C., 1 (1969).

8. A. von Engel. In *Ionized Gases*, 2nd ed. Oxford Univ. Press: London (1965).

9. J. D. Thornton. In *Chemical Reactions in Electrical Discharges*, ed. B. D. Blaustein. Advances in Chemistry Ser., *No. 80*. Amer. Chem. Soc.: Washington, D.C., 372 (1969).

10. A. V. Phelps. Proc. Conf. Phys. Quantum Electronics, ed. P. Kelley. McGraw-Hill: New York, p. 546 (1966).

11. G. Francis. In *Handbuch der Physik*, Vol. 22, ed. S. Flugge. Springer Verlag: Berlin (1956).

12. T. Maronne. *Phys. Rev., 141*, 27 (1966).

13. J. T. Massey and S. M. Cannon. *J. Appl. Phys., 36*, 361 (1965).

14. J. E. Morgan, L. F. Phillips, and H. I. Schiff. *Discuss. Farad. Soc., 33*, 119 (1962).

15. S. C. Brown. *Introduction to Electrical Discharges in Gases*. Wiley: New York (1966).

16. H. Kobayashi, M. Shen, and A. T. Bell. Res. Contr., AD–762, 480, Office of Naval Res. (15 June 1973).

17. A. T. Bell. *I & EC Fund., 11*, 209 (1972).

18. A. T. Bell and K. Kwong. *I & EC Fund., 12*, 90 (1973).

19. H. Yasuda and C. E. Lamaze. *J. Appl. Polym. Sci., 17*, 1519 (1973).

20. H. Yasuda and C. E. Lamaze. *J. Appl. Polym. Sci., 17*, 1533 (1973).

21. H. Yasuda and T. Hirotsu. *J. Appl. Polym. Sci., 22*, 1195 (1978).

22. T. Williams and M. W. Hayes. *Nature*, London, *209*, 769 (1966).

23. A. R. Westwood. *Europ. Polym. J., 7*, 361 (1971).

24. A. W. Adamson. *Phys. Chem. of Surfaces*. Interscience: New York, p. 479 (1960).

25. A. R. Denaro, P. A. Owens, and A. Crashaw. *Europ. Polym. J., 4*, 93 (1968).

26. J. M. Halovka. Sandia Labs., Res. Rept. No. SLA 74–0181 (1974).

27. Inficon Film Thickness Monitor, Model 321.

28. H. Melville and B. G. Gowenlock. *Experimental Methods in Gas Reactions*. Macmillan: London (1964).

29. H. Yasuda, O. Baumgarner, and J. J. Hilman. *J. Appl. Polym. Sci., 19*, 531 (1975); *19*, 1403 (1975).

30. H. Yasuda and T. Hirotsu. *J. Polym. Sci., Polym. Chem. Ed., 16*, 743 (1978).

31. A. R. Westwood. *Europ. Polym. J., 7*, 363 (1971).

32. K. C. Brown. *Europ. Polym. J., 8*, 117 (1972).

33. H. Yasuda, M. O. Baumgarner, H. C. Marsh, and N. Morosoff. *J. Polym. Sci., Polym. Chem. Ed., 14*, 195 (1976).

34. H. Kobayashi, M. Shen, and A. T. Bell. *J. Macromol. Sci., Chem. Ed. A8*(8), 1345 (1974).

35. K. C. Brown. *Europ. Polym. J.*, *8*, 117 (1972).
36. P. M. Hayes. In *Chemical Reactions in Electrical Discharges*, ed. B. D. Blaustein. Advances in Chemistry Ser., *No. 80*. Amer. Chem. Soc.: Washington, D.C., 350 (1969).
37. H. Yasuda and C. E. Lamaze. *J. Appl. Polym. Sci.*, *15*, 227 (1971).
38. H. Yasuda and T. Hirotsu. *J. Polym. Sci., Polym. Chem. Ed.*, *16*, 229 (1978).
39. L. F. Thompson and K. G. Mayhan. *J. Appl. Polym. Sci.*, *16*, 2291 (1972).
40. L. F. Thompson and K. G. Mayhan. *J. Appl. Polym. Sci.*, *16*, 2317 (1972).
41. H. Melville and B. G. Gowenlock. *Experimental Methods in Gas Reactions*. Macmillan: London (1964).
42. H. Kobayashi, M. Shen, and A. T. Bell. *J. Macromol. Sci., Chem.*, *A8*, 1345 (1974).
43. A. R. Denaro, P. A. Owens, and A. Crawshaw. *Europ. Polym. J.*, *5*, 471 (1969).
44. A. R. Denaro, P. A. Owens, and A. Crawshaw. *Europ. Polym. J.*, *6*, 487 (1970).
45. K. C. Brown. *Europ. Polym. J.*, *8*, 117 (1972).
46. H. Hiratsuka, G. A. Kovali, M. Shen, and A. T. Bell. *J. Appl. Polym. Sci.*, *22*, 917 (1978).
47. H. Kobayashi, A. T. Bell, and M. Shen. *Macromolecules*, *7*, 277 (1974).
48. L. Thompson and K. G. Mayhan. *J. Appl. Polym. Sci.*, *16*, 2317 (1972).
49. H. Yasuda and T. S. Hsu. *J. Poly. Sci., Polym. Chem. Ed.*, *16*, 415 (1978).
50. H. Yasuda and T. Hirotsu. *J. Polym. Sci., Polym. Chem. Ed.*, *16*, 2587 (1978).
51. H. Yasuda and T. Hirotsu. *J. Polym. Sci., Polym. Chem. Ed.*, *16*, 743 (1978).
52. M. Duval and A. Théorêt. *J. Electrochem. Soc., Solid State Sci. and Technol.*, *122*(4), 581 (1975).
53. D. D. Neiswender. In *Chemical Reactions in Electrical Discharges*, ed. B. D. Blaustein. Advances in Chemistry Ser., *No. 80*. Amer. Chem. Soc.: Washington, D.C., 338 (1969).
54. F. J. Vastola and J. P. Wightman. *J. Appl. Chem.*, *14*, 69 (1964).
55. J. K. Stille, R. L. Sung, and J. Vander Kooi. *J. Org. Chem.*, *30*, 3116 (1965).
56. H. Yasuda and T. Hirotsu. *J. Polym. Sci., Polym. Chem. Ed.*, *16*, 313 (1978).
57. N. Morosoff, W. Newton, and H. Yasuda. *J. Vac. Sci., Technol.*, *15*(6) (Nov./Dec. 1978).
58. T. W. Shaw. *J. Chem. Phys.*, *30*, 1366 (1959).
59. R. Kelley and C. A. Winkler. *Can. J. Chem.*, *37*, 62 (1959).
60. H. Yasuda and T. Hsu. *J. Polym. Sci., Polym. Chem. Ed.*, *15*, 81 (1977).

6. Structure of Plasma Polymers

1. Introduction

Polymers prepared by ultraviolet irradiation of monomer show expected differences in structure from those formed by more conventional techniques, since the source is of relatively low energy. A mercury vapor lamp emitting light of a wavelength > 2000 Å provides energy of less than 6 eV, which is insufficient to induce ionization. Furthermore, in photolysis, the partial pressure of monomer is relatively high, such as a few torr, compared with that used in electron bombardment, which may range between 10^{-5} to 10^{-2} torr if electron-gas interaction is minimized. Although electron energies can have values up to 1000 eV, films have been obtained at energies as low as 6 eV. In a glow discharge in which the pressures employed may range from 0.05 to 10 torr, the discharge is a relatively high energy process involving the formation of a variety of active species, including electrons and ions. Such energies give rise to decomposition products that are less selective than those observed in ultraviolet radiation.

Early work with glow discharges in gaseous organic monomers produced coherent films. These reactions had the characteristics of a gas-phase electrolysis, one ion fragment deposited at a time, to give a highly crosslinked material.

2. Plasma Oils

It has been shown in chapters 4 and 5 that passage of an organic vapor through a glow discharge may produce, depending on the experimental conditions, oils, films, and powder. Oils obtained from ethylene in a r.f. plasma are soluble in appropriate solvents and show generally an infrared spectrum, as illustrated in figure 6.1.

On the basis of published tabulated data the assignments of bands have been made that are shown in table 6.1. NMR spectra of 4% solutions of the

Figure 6.1. IR spectrum of oil formed by r.f. plasma polymerization of ethylene. *Source:* Adapted from J. M. Tibbitt, M. Shen, and A. T. Bell, *J. Macromol. Sci., Chem., A10*(8), 1623 (1976). Reprinted by courtesy of Marcel Dekker, Inc., N.Y.

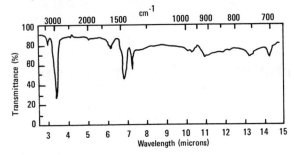

oils yielded the absorption bands shown in table 6.2. Finally, mass spectra of oil pyrolyzates were found to contain aliphatic, olefinic, and singly substituted phenyl groups. The presence of methyl, methylene, tertiary carbon, and olefinic groups in plasma oil from ethylene has been described in the literature [1,2,3].

Such plasma oils obtained from ethylene are essentially branched oligomers having low molecular weights on the order of 400–600 [1,4]. When the oil

Table 6.1. Assignments of IR absorption bands for plasma oils from ethylene

Absorption frequency (cm^{-1})	Assignment
3400	O—H stretch
3000	C—H stretch in $C=C\overset{H}{\diagup}$
2940	C—H stretch in CH_2, CH_3
2880	
1960	C=C stretch in $C=C=C$
1670	C=C stretch
1640	(nonconjugated)
1605	C=C stretch
	(conjugated)
1495	C⋯C stretch in —C_6H_5
1455	C—H bend in CH_2
1447	C—H bend in CH_3
1370	C—H bend in CH_3
990	C—H out-of-plane deformation in —C—CH_2
965	C—H out-of-plane deformation in —C=C(trans)
905	C—H out-of-plane deformation in —C=CH_2
752	C—H bend in monosubstituted phenyl group C_6H_5—R
700	

Source: J. M. Tibbitt, M. Shen, and A. T. Bell, *J. Macromol. Sci., Chem., A10*(8), 1623 (1976). Reprinted by courtesy of Marcel Dekker, Inc., N.Y.

Table 6.2. Assignment of NMR absorption band
for plasma oils from ethylene

Chemical shift (ppm)	Assignment
0.96	$R-CH_3$
1.26	$R-CH_2-R$
1.61	R_2-CH-R
1.91	$-C=C-CH_2-R$
2.31	$C_6H_5CH_2-R$
5.01 5.26 5.41	Various $\begin{matrix} \diagdown \\ \diagup \end{matrix} C=C \begin{matrix} \diagup \\ \diagdown \\ H \end{matrix}$
6.14	$C=C-\overset{\overset{\displaystyle H}{\mid}}{C}=C-$ with lower H
7.04	C_6H_5-H

Source: J. M. Tibbitt, M. Shen, and A. T. Bell, *J. Macromol. Sci., Chem., A10*(8), 1623 (1976). Reprinted by courtesy of Marcel Dekker, Inc., N.Y.

was dissolved in *m*-Xylene after removal from the reactor and stored at $-10°C$ for one week, however, the molecular weight was found to have increased to nearly 1500. This observation suggested the presence in the oil of reactive species that caused the polymerization to continue. Electron-spin resonance measurements of the freshly deposited oil indeed confirmed the presence of free radicals, the concentration of which decreased by about 50% within five days. Infrared spectra of the oil deposition on sodium chloride substrate indicated that continued exposure to air caused oxidation of reactive sites to form hydroxy- or carboxy-type functional groups [4].

In order to obtain further information on the structure of plasma polymerized hydrocarbons, Bell and colleagues [5] conducted an extensive analysis of these materials by the use of a pyrolysis/gas chromatography (P/GC) technique [6]. On the basis of energies of dissociation, bond rupture at points α and β to tertiary carbon atoms is the preferred scission mechanism. If hydrogenation follows pyrolysis, bond breakage at these points suggests that the fragmentation products contain isoalkanes. High levels of isoalkanes are taken as indicative for sizable branching and crosslinking in the original polymer. Therefore, the isoalkene content in the G/PC output would represent a measure of the concentration and length of branches and crosslinks in the parent polymer.

The authors [5] observed that the P/GC data are in qualitative agreement with their NMR results (table 6.2), in that the extent of isoalkane content correlates with the concentration of tertiary carbon atoms. Figures 6.2 and 6.3 represent pyrograms of the oil, film, and powder products obtained from r.f. glow-discharge polymerization of ethylene and butadiene, respectively. It is noted that in both cases, the isoalkane content increases in going from oil to film to powder.

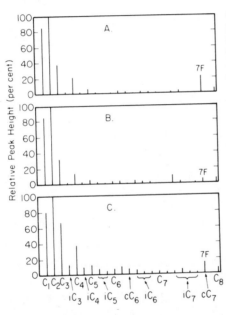

Figure 6.2. Pyrograms derived from plasma-polymerized ethylene samples: (A) oil, (B) film, (C) powder. *Source:* M. Seeger, R. J. Gritter, J. M. Tibbitt, M. Shen, and A. T. Bell, *J. Polym. Sci., Polym. Chem. Ed.,* *15,* 1403 (1977). Copyright © 1977, John Wiley & Sons, Inc. Reprinted by permission of John Wiley & Sons, Inc.

It was mentioned above that in an r.f. plasma (13.56 MHz), a high concentration of isoalkanes in the pyrolysis products reflects a high density of crosslinks and/or branches in the pyrolyzed material. Another observation of interest is that the 7F fragment, which reflects the presence of aromatic groups, is much stronger in the plasma oil than in the corresponding plasma film. In the combined employment of infrared spectroscopy and NMR [7], increased branching and crosslinking yet lower aromatic content in film samples compared with oil products have also been detected. These results are listed in table 6.3. On the other hand, when hydrocarbons such as ethylene, butadiene, and pentane were plasma-polymerized under a.c. (20 KHz, 250 V,

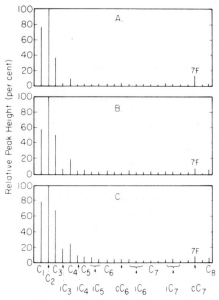

Figure 6.3. Pyrograms derived from plasma-polymerized butadiene samples: (*A*) oil, (*B*) film, (*C*) powder. *Source:* M. Seeger, R. J. Gritter, J. M. Tibbitt, M. Shen, and A. T. Bell, *J. Polym. Sci., Polym. Chem. Ed., 15*, 1403 (1977). Copyright © 1977, John Wiley & Sons, Inc. Reprinted by permission of John Wiley & Sons, Inc.

10–20 mA) conditions, no aromatic groups were observed in the polymer formed [8].

Analysis by pyrolysis/gas chromatography of oil obtained from both r.f. and a.c. glow discharges of benzene disclosed that a large proportion of the fragments were composed of aromatic compounds including toluene, the xylenes, ethylbenzene, naphthalene, and polyalkylbenzenes. Mass spectra confirmed that the cyclic products detected in the pyrograms were derived from structures in the original plasma polymer rather than from cyclization

Table 6.3. Functional group characterization of oils and films obtained by plasma polymerization of ethylene and butadiene

Monomer	Polymer form	Functional group m mole/gm polymer		Apparent cross-link density[a]	Isoalkane (% of C_2)	7F Relative peak height (% of C_2)
		[R_3CH]	[C_6H_5]			
Ethylene	oil	9.5	1.2	∞	27	12
	film	18.0	0.39	6.4	36	6
Butadiene	oil	5.9	2.6	∞	15	22
	film	12.0	12.0	9.6	28	3

[a]Backbone carbons per crosslink.
Source: M. Seeger, R. J. Gritter, J. M. Tibbitt, M. Shen, and A. T. Bell, *J. Polym. Sci., Polym. Chem. Ed., 15*, 1403 (1977). Copyright © 1977, John Wiley & Sons, Inc. Reprinted by permission of John Wiley & Sons, Inc.

Table 6.4. Functional group concentrations in hydrocarbon oils produced by plasma polymerization

Monomer	[RCH₃]	[R₂CH₂]	[R₃CH]	[C=C]	[C₆H₅]	[CH₃]/[C₆H₅] IR	[CH₃]/[C₆H₅] NMR
Ethylene	8.3	17.0	9.5	3.5	1.2	6.9	6.6
Ethylene-acetylene	8.9	17.0	10.0	3.5	1.2	7.5	7.3
Butadiene	5.6	15.0	5.9	4.5	2.6	2.1	1.9
Benzene	5.9	19.0	18.0	6.1	11.0	0.51	0.59

Source: J. M. Tibbitt, M. Shen, and A. T. Bell, *J. Macromol. Sci., Chem., A10*(8), 1623 (1976). Reprinted by courtesy of Marcel Dekker, Inc., N.Y.

caused by pyrolysis. The functional group concentrations detected in plasma oil obtained in a r.f. glow discharge are listed in table 6.4.

Bell and his coworkers [1] postulated the molecular structure of the oil molecules (see figure 6.4). The authors point out that this structure may not be complete, since no account has been taken on the possible existence of aliphatic rings or quaternary carbon atoms, both of which should be present to some extent. It is again noted that aromatic groups were not detected in such oils when they formed under a.c. discharge conditions. In general, the structures of polymerized hydrocarbons are similar to each other but bear little resemblance to their commercial polymer counterparts. Rather, it is a complex

Figure 6.4. Structure of oil obtained from ethylene in a r.f. glow discharge as postulated by Bell et al. *Source:* J. M. Tibbitt, M. Shen, and A. T. Bell, *J. Macromol. Sci., Chem. A10*(8), 1623 (1976). Reprinted by courtesy of Marcel Dekker, Inc., N.Y.

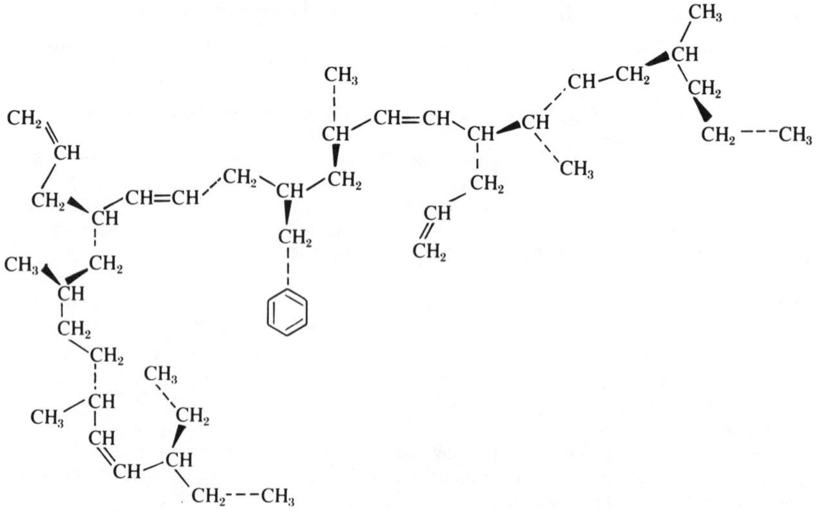

arrangement of short-chain segments, branches, crosslinks, double bonds, and aliphatic and aromatic rings.

3. Plasma Films

A. Microstructure

As a consequence of the high degree of crosslinking and perhaps also the presence of oxygen in glow-discharge polymers, the density of these materials is generally found to be higher than that of conventionally polymerized counterparts. In fact, the density of plasma polymers has been shown to exceed even the crystal density of the corresponding linear polymers (table 6.5) [9]. Low-angle X-ray diffraction patterns indicate that plasma polymers are amorphous; this would be expected from the aforementioned complex arrangement of a variety of structural entities in plasma materials. For any given monomer the density of the plasma polymerized material varies with the operating conditions; therefore a range of the measured densities of plasma polymers is given in table 6.5.

Using infrared spectroscopy and NMR, the concentration of the functional groups could be determined [7], and from this data an apparent crosslinking density of each polymer could be calculated [5]. Table 6.6 summarizes the concentrations of tertiary carbon atoms and phenyl groups and the calculated crosslinking densities, as well as the sum of the height of all isoalkane peaks and the height of the 7F peak, characteristic of aromatic fragments.

These results imply that polymers formed by glow discharges should be hydrogen-deficient when compared with their respective monomers. This has been experimentally confirmed in a study in which a series of unsaturated and saturated hydrocarbons were plasma-polymerized under the condition of 2 torr pressure, 5 cm^3/min flow rate, and 100 watts, at a radio frequency of 13.56 MHz (table 6.7) [10].

Table 6.5. Polymer densities

Starting monomer	Plasma polymer (g/cm^3)	Crystal of linear polymer (g/cm^3)	Amorphous linear polymer (g/cm^3)
Styrene	1.332–1.408	1.126	1.04
Ethylene	1.141–1.231	1.014	0.852
Vinyl chloride	1.463–1.499	1.44–1.522	1.39
Vinyl fluoride	1.341–1.395	1.44	—
Benzene	1.331–1.364	—	—

Source: W. W. Knickmeyer, B. W. Peace, and K. G. Mayhan, *J. Appl. Polym. Sci., 18,* 301 (1974). Copyright © 1974, John Wiley & Sons, Inc. Reprinted by permission of John Wiley & Sons, Inc.

Table 6.6. Functional group characterization of plasma polymerized ethylene, butadiene, and benzene

Monomer	Functional group (m mole/gm polymer)		Apparent cross-link density[a]	Isoalkane (% of C_2)	7F Relative peak height (% of C_2)
	$[R_3CH]$	$[C_6H_5]$			
Ethylene	18.0	0.39	6.4	82	10
Butadiene	12.0	1.20	9.6	66	20
Benzene	26.0	7.90	6.9	20	83

[a]Backbone carbons per crosslink.
Source: M. Seeger, R. J. Gritter, J. M. Tibbitt, M. Shen, and A. T. Bell, *J. Polym. Sci., Polym. Chem. Ed., 15,* 1403 (1977). Copyright © 1977, John Wiley & Sons, Inc. Reprinted by permission of John Wiley & Sons, Inc.

The last column in table 6.7 lists values for the H/C ratio for each polymer divided by that of the monomer. It can be seen that each of the polymers is hydrogen-deficient with respect to its monomer. Acetylene appears to be unique in this regard, since its polymer has nearly the same number of hydrogen atoms per carbon atom as has the monomer. A second group of polymers that can be distinguished in table 6.7 contains approximately 10–30% hydrogen deficiency and is derived from double-bond-containing monomers. Interestingly, polymers derived from saturated hydrocarbons are found to be the most hydrogen-deficient, having lost nearly half of their hydrogen atoms in the polymerization process.

Hydrogen deficiency of plasma polymers as compared with their respective conventional polymer species represents another piece of evidence of the highly crosslinked three-dimensional plasma network containing considerable amounts of unsaturation. Still further confirmation of these characteristics was

Table 6.7. Carbon-hydrogen ratios of plasma-polymerized hydrocarbons

Compound	H/C Ratio		H/C (Polymer)
	Polymer	Monomer	H/C (Monomer)
Acetylene	0.95	1.00	0.95
Ethylene	1.49	2.00	0.75
Propylene	1.40	2.00	0.70
Isobutylene	1.44	2.00	0.72
Cis-2-Butene	1.34	2.00	0.67
Butadiene	1.33	1.50	0.88
Methane	2.40	4.00	0.60
Ethane	1.55	3.00	0.52
Propane	1.58	2.67	0.59

Source: H. Kobayashi, A. T. Bell, and M. Shen, Office of Naval Res., Proj. No. NR 256–526 (Dec. 1, 1973).

obtained from solvent swelling data. It is well known that the degree of swelling decreases with increasing crosslink concentration in the polymer until swelling is no longer detected in very highly crosslinked materials. Attempts to dissolve or swell rigid polymer films obtained from plasma polymerization of hydrocarbons have indeed been unsuccessful [10].

The pyrograms in figures 6.2 and 6.3 revealed that the concentrations of isoalkane fragments increase from oil to film to powder. Furthermore, the 7F fragment, which indicates the presence of aromatic groups, is considerably stronger in the oil than in the film. On the other hand, branching in the films is increased, and high crosslinking occurs. Figure 6.5 compares pyrograms of r.f. plasma-polymerized ethylene (PPE), acetylene (PPA), butadiene (PPBd), and benzene (PPB) [5]. The pyrograms of PPE and PPBd are similar, the former showing a somewhat lower isoalkane content; PPA and PPB display even lower isoalkene concentrations than PPE. PPA exhibits a particularly strong methane peak, which implies very high crosslink density and unsaturation in the parent polymer. The benzene pyrogram shows two strong peaks, 6B (cyclohexane) and 7F (methylcyclohexane), which imply a high degree of aromaticity in the parent polymer. The extent of these peaks is smaller in PPBd and PPE, reflecting a lower aromatic content in plasma-polymerized

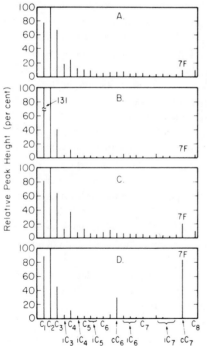

Figure 6.5. Pyrograms derived from plasma-polymerized hydrocarbons: (*A*) ethylene, (*B*) acetylene, (*C*) butadiene, (*D*) benzene. *Source:* M. Seeger, R. J. Gritter, J. M. Tibbitt, M. Shen, and A. T. Bell, *J. Polym. Sci., Polym. Chem. Ed., 15,* 1403 (1977). Copyright © 1977, John Wiley & Sons, Inc. Reprinted by permission of John Wiley & Sons, Inc.

butadiene and ethylene. It appears that the pyrolysis/gas chromatography technique (P/GC), which has been demonstrated by Bell and colleagues [5] to yield the aforementioned important data, will play a significant role in the development of a detailed description of the microstructure of plasma polymers.

Recently, Kronick and colleagues [2] surveyed the infrared spectra of glow-discharge polymers (200–300 V a.c., 20 KHz, 1 torr) from several monosubstituted benzenes. Prominent features of their spectra are listed as infrared groups in table 6.8. Under such operating conditions aromatization does not occur; therefore all the aromatic structures listed derive most likely from molecules of the starting vapor. Details of the aromatic substitution are difficult to ascertain from infrared spectra alone. The regions between 1650 and 2000 cm^{-1} were obscured by an intense band assigned to a CO bond. Resolved bands were observed at 820, 755, 725, and 695 cm^{-1}; meta-substitution could give bands near 820 and 690 cm^{-1}, para-substitution near 820 cm^{-1}, and monosubstitution near 690 and 755 cm^{-1}. All spectra in this study [2] show a band at 755 cm^{-1} with the same intensity relative to that at 690 cm^{-1}.

It has frequently been observed, as in the above study [2], that the chemical processes that occur in a low-frequency glow discharge are different from those in an r.f. discharge. Furthermore, evidence has been presented [11] that the "T-spectrum" in the emissions from the positive columns of helium discharge with admixtures of benzene, toluene, and chlorobenzene is due to a small aliphatic fragment, probably C$_2$H or C$_4$H$_x$. This spectrum is not observed in the positive column with carrier gases of lower ionization energy than helium, for example, argon or neon, nor in aromatic vapor alone. Yet, the T-spectrum from discharges of pure toluene or phenylacetylene was observed in the negative glow and cathode glow [2]. Consistent with this finding, the polymers from these vapors have acetylenic infrared bands at 2100, 2200, and 3300 cm^{-1}. Since plasma films produced in the negative glow, where the maximum ion energies are lower, have chemical features similar to those on the cathode, but greater solubility, the authors concluded [2] that ions with high kinetic energy do not play an important role in determining detailed chemical structure, except for crosslinking. They also concluded that the process of plasma polymerization of monosubstituted benzenes is largely one of fragmentation of aromatic rings to form branched polymer chains with possible subsequent addition of starting molecules with elimination of a substituent. Some aromatic material does react directly, as evidenced by weaker bands in the infrared spectrum of the plasma films at 820 cm^{-1} (para-substitution) and at 775 cm^{-1} (1,3-substitution) [12].

In comparing the behavior of hydrocarbons with that of fluorocarbons in an electrodeless glow discharge, Yasuda and Hsu [13,14] reported a distinct

Table 6.8. Group frequencies found in plasma films from monosubstituted benzenes

	Frequency (cm^{-1})						
	Chlorobenzene	Fluorobenzene	Benzonitrile	Acetophenone	Phenylacetylene	Nitrobenzene	Toluene
Aromatic							
$\nu(=CH)$	3050 3020		3050	3050 3015	3050 3015	3050	3050 3020
$\nu(C=C)$	1590	3020 1600 1580 1500 1480	1590	1595 1575	1595 1575	1585	1595
$\beta(=CH)$	1490 1150 1070 1025	1480	1480 1150 1060 1010	1490 1150 1065 1020	1490 1150 1065 1020	1480 1150 1060 1010	1490 1150 1070 1030
$\gamma(=CH)$	820		780				
$\delta'(ring)$	755 690		750 680	750 690	750 690	740 690	740 690
Aliphatic							
$\gamma(CH)$	960 840		950 830	950 835	960 835	950	960 840
$\nu_{as}(CH_2)$	2915	2915	2915	2915	2915	2915	2915
$\nu_s(CH_2)$	2850	2850	2850	2850	2850	2850	2850
$\nu_{as}(CH_3)$	Absent	Absent	Absent	2950	2950		2950
$\nu_s(CH_3)$	Absent	Absent	Absent	2880	2880		2880
$\delta_s(CH_3)$	Absent	Absent	Absent	1370	1370		1370

Source: P. L. Kronick, K. F. Jesch, and J. E. Bloor, *J. Polym. Sci., A-1(7)*, 767 (1969). Copyright © 1969, John Wiley & Sons, Inc. Reprinted by permission of John Wiley & Sons, Inc.

difference between these two monomer categories. Tetrafluoroethylene polymerization was found to compete with simultaneous decomposition of the monomer, while polymerization of ethylene proceeded without decomposition under similar conditions. The detachment of hydrogen was not considered as decomposition in this context.

Results from electron spectroscopy for chemical analysis (ESCA) indicate that the plasma polymer of tetrafluoroethylene formed in an incomplete glow region (high flow rate and low discharge power) is a hybrid of polymers of plasma polymerization and plasma-induced polymerization. Polymers formed under "high glow" conditions are alike regardless of the extent of decomposition of the monomer. The typical plasma polymers of tetrafluoroethylene contain carbons with different numbers of F, unlike Teflon.

The chemical structure of r.f. glow-discharge polymers from halogenated vinyl compounds has been investigated by Westwood [15]. The results of the chemical analysis of these materials are listed in table 6.9. Their infrared spectra show the following characteristics. The vinylidene fluoride polymer (PVDF) shows two strong bands at 1271 and 1150 cm^{-1}, which arise from the symmetric and antisymmetric stretching vibration in CF_2 groups. This type of vibration is particularly sensitive to changes of internal rotation. Therefore, lack of any regular chain conformation can readily account for the slight shift in these bands from their frequencies in conventional PVDF, at 1274 and 1180 cm^{-1}. The presence of methylene groups is indicated by the CH_2 bands at 2970 cm^{-1} and bands at 1446 and 1393 cm^{-1}. The bands that arise from carbon-chain skeletal vibrations are absent; the inference is that the polymer contains no long chain sections and must be crosslinked.

The presence of carbonyl vibration at 1760 cm^{-1}, indicating either aldehyde or ketone groups, is presumably due to oxygen uptake upon exposure of the film to air. The band at 940 cm^{-1} is assigned to bending vibrations in ethylenic structures. The low C/H ratio in table 6.8 points to considerable unsaturation and crosslinking.

Table 6.9. Composition of plasma polymers from vinyl halides

Polymer	Discharge conditions	Empirical formula
PVC	116 V/cm, 1.75 mA/cm², 0.76 torr	$C_2H_{2.06}Cl_{0.51}O_{0.40}$
PVC (vinyl chloride/argon)	240 V/cm, 2.95 mA/cm², 1.14 torr	$C_2H_{1.97}Cl_{0.57}O_{0.28}$
PVF	148 V/cm, 2.25 mA/cm², 1.52 torr	$C_2H_{2.32}F_{0.31}O_{0.16}$
PVF	113 V/cm, 1.95 mA/cm², 0.57 torr	$C_2H_{2.25}F_{0.24}O_{0.37}$
PVDF	100 V/cm, 1.72 mA/cm², 0.88 torr	$C_2H_{1.45}F_{0.95}O_{0.15}$
PVDF (powder form)	100 V/cm, 1.72 mA/cm², 0.88 torr	$C_2H_{1.46}F_{0.98}O_{0.16}$
PVDF	109 V/cm, 1.85 mA/cm², 0.95 torr	$C_2H_{1.69}F_{0.99}O_{0.07}$

Source: A. R. Westwood, *Europ. Polym. J.*, *7*, 377 (1971). Reprinted with permission from *European Polymer Journal.* Copyright © 1971, Pergamon Press, Ltd.

In contrast to commercial polyvinylfluoride, PVF, the plasma analogue shows a lower fluorine content, some oxygen, and considerable unsaturation (figure 6.6). The low fluorine content is confirmed by relatively weak C—F stretching bands in the IR. As shown in the PVDP (figure 6.7), the carbonyl band occurs again. The shift in frequency, when compared to the plasma PVDP, reflects the much lower fluorine content. Two bands are assigned to C=C stretching vibrations. The first, at 1680 cm^{-1}, may arise from F—C groups

$$
\begin{array}{cc}
\text{F} & \text{F} \\
| & | \\
-\text{C}=\text{C}-
\end{array}
$$

The other, at 1634 cm^{-1}, may be due to unsaturated groups that do not contain fluorine.

The analytical data of plasma-polymerized vinyl chloride corresponds to the empirical formula of $C_{10}H_{10} Cl_{2.5}O_2$. The CH_2 deformation vibration is shifted to a higher frequency, 1448 cm^{-1}, as compared with conventional PVC, indicating some chlorine loss (figure 6.8).

In general, results from infrared spectroscopy and chemical analysis of plasma polymers from vinyl halides indicate that the materials are crosslinked

Figure 6.6. IR spectrum of plasma polymer from vinylfluoride, 2.25 mA/cm^{-2}; 1.52 torr. Empirical formula—$C_2H_{2.32}F_{0.31}O_{0.16}$. *Source:* Adapted from A. R. Westwood, *Europ. Polym. J., 7,* 377 (1971). Reprinted with permission from *Europ. Polym. J.* Copyright 1971, Pergamon Press, Ltd.

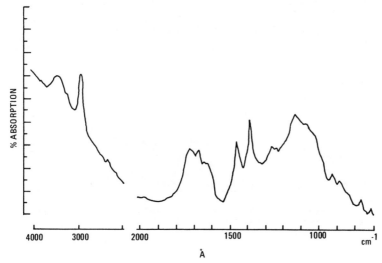

Figure 6.7. IR spectrum of plasma polymer from vinylidene fluoride. 1.72 mA/cm^{-2}; 0.88 torr. Empirical formula—$C_2H_{1.45}F_{0.95}O_{0.15}$. *Source:* Adapted from A. R. Westwood, *Europ. Polym. J., 7,* 377 (1971). Reprinted with permission from *Europ. Polym. J.* Copyright 1971, Pergamon Press, Ltd.

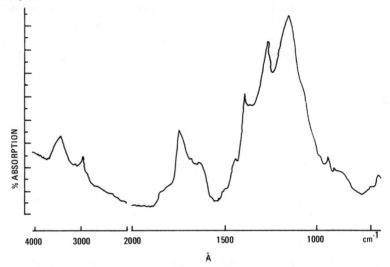

Figure 6.8. IR spectrum of plasma polymer from vinyl chloride. 1.75 mA/cm^{-2}; 0.76 torr. Empirical formula—$C_2H_{2.06}Cl_{0.51}O_{0.40}$. *Source:* Adapted from A. R. Westwood, *Europ. Polym. J., 7,* 377 (1971). Reprinted with permission from *Europ. Polym. J.* Copyright 1971, Pergamon Press, Ltd.

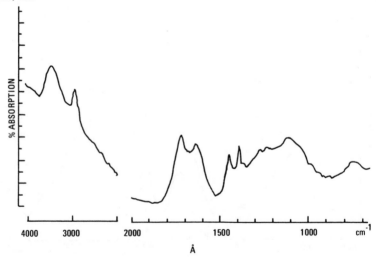

aliphatic structures. They all contain double bonds and add oxygen on exposure to air. This latter tendency, observed in most plasma polymers, is known to be due to the presence in the polymer of residual free radicals. Vinylidene fluoride exhibited a tendency to form polymer in the gas phase. For this plasma polymer, the structure below, with crosslinks to neighboring chains, was suggested [15].

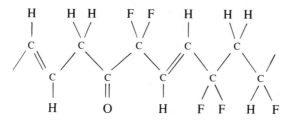

The oxygen effect is largely confined to the surface regions [16], indicating that the reaction of the polymer film with atmospheric oxygen is initiated rapidly at the surface. The crosslinked nature of the film then permits only slow diffusion into the subsurface and, eventually, into the bulk. This mechanism is supported by IR studies of powdered samples of the polymer prepared in the freestanding vacuum-line apparatus. The oxygen function is detected for the high-surface-area fine powder by absorptions in the OH region and carbonyl regions. The IR data from the samples exposed to atmosphere suggest that the oxygen functionality is also present in the form of carboxyl groups. It is generally found that the oxygen functionality originates in the reaction of molecular oxygen with unsaturated/radical sites at the surface.

It is noted that under the inductively coupled r.f. conditions, 0.25 torr, 4 MHz, 22 W, polymer obtained from vinyl acetate revealed very little unsaturation in its IR spectrum and showed considerable swelling in methyl ethyl ketone [7]. The rate of conversion was lower than those of styrene or vinyl chloride. On the other hand, acrylonitrile formed a polymer very rapidly under these experimental conditions. At high input rates, a powder formed that was partially soluble in dimethyl formamide. Although the insoluble film gave a very diffuse infrared spectrum, it was very similar to a conventional polyacrylonitrile spectrum.

Perfluorobutene-2 was readily converted into solid plasma films, at 13.56 MHz, 70 W, 0.5 torr. The properties of the film support a branched fluorocarbon structure [17]. Strong electron-spin resonance signals observed in the film indicate the presence of free radicals, a common characteristic of films formed in plasma.

It has been shown in chapter 5, figure 5.17, that the addition of 10 cm³/min of dichlorodifluoroethane to a flow of 50 cm³/min of methane in a r.f. discharge increases the polymer deposition rate fiftyfold over that for methane alone, while the addition of CF_4 has only a negligible effect on the rate. As a test of the extent to which the halogenated additives polymerized by themselves in a discharge at 0.5 torr and 100 W, 13.56 MHz, several methyl halides were subjected to glow discharge polymerization, but only methylene chloride, CH_3Cl, was found to polymerize at an appreciable rate. The results suggested [18] that only CH_3Cl acts as both a catalyst and a monomer. Representative compositions of the polymers formed, when a halogenated compound is added to methane, are shown in table 6.10.

It is evident that the use of either CF_2Cl_2 or CH_3Cl causes a negligibly small amount of chlorine to be incorporated into the polymer. Furthermore, the H/C ratio of the polymer prepared from a mixture of methane and CH_3Cl or CF_2Cl_2 is considerably smaller than that of a polymer prepared from methane alone.

B. Morphology of Plasma-Deposited Films

A study of the submicroscopic structure of plasma-polymerized films presents intriguing aspects that appear to be related to both the mechanism of film formation and the properties of the polymer formed. In 1974, Bell and his coworkers [19], when investigating the effects of substrate surface roughness on the morphology of plasma-polymerized ethylene, noted in electron micrographs the gradual formation of spherical polymer particles on an electropolished aluminum surface.

After a 1 min deposition time, spheres of about 0.2 μ in diameter were observed. After a 2–3 min deposition time, the size of the spheres was slightly larger, that is, 0.2–0.3 μ in diameter. The substrate surface also contained

Table 6.10. Composition of plasma polymerized polymer (2 torr, 100 W, 13.56 MHz)

Monomer	Flow rate (cm³/min)		Empirical formula of polymer film
CH_4	5.0		$CH_{2.40}$
CH_4	50	mixture	$CH_{1.62}Cl_{0.008}$
CF_2Cl_2	2.3		
CH_4	50	mixture	$CH_{1.49}Cl_{0.02}$
CH_3Cl	6.4		
CH_3Cl	6.4		$CH_{0.80}Cl_{0.46}$

Source: L. F. Thompson and K. G. Mayhan, *J. Appl. Polym. Sci., 16,* 2291 (1972). Copyright © 1972, John Wiley & Sons, Inc. Reprinted by permission of John Wiley & Sons, Inc.

irregularly shaped polymer lumps ranging in diameter from 400–1000 Å. These lumps appeared to increase in size until their growth was stopped by coalescence. For longer polymerization times, the film became smoother as more polymer was deposited between the coalesced lumps. The larger spherical shapes continued to grow as the deposition proceeded. Their diameter was about 0.5 μ after 10 min of polymerization and about 1.5 μ at 30 min. Since these features are always found on the surface of a polymer film, the authors suggested that they are due to spherical particles grown in the gas phase and that they descended to the substrate surface at the end of the polymerization period.

At about the same time, other investigators examined plasma-formed polymers with scanning electron microscopy (SEM) and observed spherical particles on the surfaces. Thompson [20] noted plasma-formed polystyrene spheres with diameters of 0.2–1.7 microns that varied in size with the power to the reactor. Depending on the reactor conditions, spherical particles were found by other investigators to range in diameter from about 0.2 to 2 microns. Suggestions were also made in the literature that this phenomenon is separate from film-forming processes, since a gradual growth was not visible by the SEM techniques.

James and colleagues [21] pointed out more recently, however, that the sphere size corresponded roughly to the smallest resolution limits of the SEM instruments employed. Using the low-angle X-ray technique, they found that the scattering curves of conventional polyethylene are completely different from those of plasma-deposited polyethylene. The investigators calculated for the latter a sphere diameter of 0.26 microns under the conditions used. For plasma-formed polystyrene, the calculated diameters were 0.19 and 0.43 micron for plasmas of 14 and 26 W.

Their calculations from X-ray analysis indicated that approximately 70% of the film volume is made up of spheres of a higher density than that of the surrounding polymer phase. Subsequently employing both transmission electron microscopy (TEM) and scanning electron microscopy (SEM) [22], these workers confirmed the existence of a two-phase structure of spheres embedded in a polymer binder in plasma-deposited polyethylene and polystyrene. A problem existed in the interpretation of the TEM micrographs. There is little doubt that the electron beam (100 KV) attacked the polymer, since the TEM samples were not metallized. The TEM micrographs still provided evidence of spherical particles, however, despite thermal attack on the polymer. James and colleagues [21] proposed the following explanation.

Thermal gravimetric analysis (TGA) has shown that plasma-formed polystyrene films exhibit higher thermal stability than their conventional counterparts, with about 40% of the original mass remaining at temperatures of 700°C

[23]. The residue was a carbon matrix form consistent with the crosslinked plasma-formed polymer structure. In contrast, conventional polystyrene has been shown [21] thermally to degrade completely at temperatures of 600°C, while plasma-deposited polyethylene was found to retain 50% of the original mass at 450°C. Evidently, plasma-formed polymers have a greater thermal stability than their conventional analogues, presumably because of the high degree of crosslinks in their structure. James and his coworkers concluded, therefore, that in the TEM beam a burning away of the lower-density material occurred while carbonizing the higher-density spheres, thus leaving a TEM-resolvable structure.

4. Plasma Powder

In 1972, Liepins and Sakaoku [24] were the first to report the generation of powder only in a glow discharge. They cite four main variables for an efficient powder formation, namely:

Type of monomer.
Monomer partial pressure in the system.
Type of inert gas used.
Design of the polymerization chamber.

Although powder formation has been observed in reactors of many different designs, these authors suggest that certain designs are more appropriate than others in generating large amounts of powder. Among three differently designed reaction systems, including a standard Bell jar, the design shown in figure 6.9 was found to be most effective in generating powder in a r.f. induction field.

It is significant that these workers operated with a total pressure in the reactor of 0.80–3.0 torr. They pointed out that a pressure above 0.6 torr was required to form powder exclusively. The upper vapor pressure was a point at which the plasma tended to be extinguished in their system. Interestingly, the powder formation was increased further by vertically filling the chamber with pieces of glass tubing. The authors suggest that the increased powder formation in this situation is due to enhanced plasma convection by the circulation cells that are formed in the glass tubings. The large amount of powder formed at the base of most of the tubings would seem to support this view.

Liepins and Sakaoku [24] also noted the importance of inert gas on the amount of powder formed; the rate of formation decreased in the following order:

Helium > Nitrogen > Neon > Argon > Air

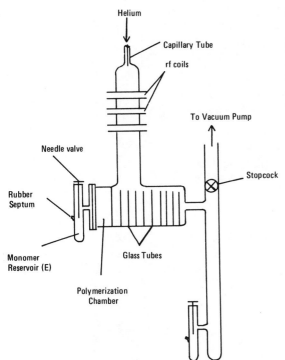

Helium

Capillary Tube

rf coils

To Vacuum Pump

Needle valve

Stopcock

Rubber
Septum

Monomer
Reservoir (E)

Glass Tubes

Polymerization
Chamber

Figure 6.9. Schematic system arrangement for powder formation. *Source:* Adapted from R. Liepins and K. Sakaoku, *J. Appl. Polym. Sci.,* *16*, 2633 (1972). Copyright © 1972, John Wiley & Sons, Inc. Reprinted by permission of John Wiley & Sons, Inc.

The data, except for nitrogen, suggest that a Penning ionization (chapter 4, section 2A) mechanism operates in the initiation of this polymerization.

Among the monomers investigated, the aromatic hydrocarbons, such as styrene, toluene, benzene, and *p*-Xylene, were most efficient in conversion to powder. The only alkane investigated in this series of monomers, hexane, was found to be the most difficult to convert into polymeric powder. The solubility of all the powders in tetrahydrofuran decreased with increasing polymerization time. These authors reported that the solubility of the powders appears to be extremely sensitive to the exact experimental conditions existing in the reactor (table 6.11). Measurements of the inherent viscosity in tetrahydrofuran disclosed that regardless of molecular structure, all soluble fractions contained polymers of very low molecular weight.

Infrared spectroscopy revealed the following features of these plasma polymers [15].

Polystyrene: All of the absorptions of polystyrene were present plus a new absorption as a shoulder at 1700–1600 cm^{-1}, possibly due to aliphatic C=C stretching vibration, or α, β-unsaturated ketone carbonyl.

Table 6.11. Monomers polymerized into powder

Monomer	Amount monomer used (g)	Amount powder (g)	Conversion (%)	Polymerization time (min)	inh[a]	Solubility (%)[b]	Color
Styrene	2.10	0.38	18	13	0.06	90	light tan
Toluene	2.60	0.39	15	16	0.05	80	light tan
Benzene	2.41	0.43	18	28	0.03	90	tan
p-Xylene	2.95	0.41	14	20	0.05	50	light tan
Hexane	5.20	0.17	3	35	0.03	60	light tan
Isoprene	2.30	0.31	13	25	—	insoluble	tan
Acetonitrile	3.05	0.40	13	20	0.04	80	dark tan
Vinyl chloride	—	0.21	—	30	0.04	80	dark brown
Tetrabutyltin	2.00	0.19	10	25	0.02	70	tan
Styrene/divinyl-benzene[c]	3.10	0.36	12	20	—	insoluble	light tan
Styrene/1,2-dibromo-ethane[c]	2.90	0.29	10	16	0.04	90	brown

[a]Determined on solutions from 0.123 to 0.315 g/100 ml of tetrahydrofuran at 30.0°C.
[b]In tetrahydrofuran; the data represent the highest solubility observed during the first 5–10 min of polymerization.
[c]A 1:1 mixture (by weight).

Source: R. Liepins and K. Sakaoku, J. Appl. Polym. Sci., 16, 2633 (1972). Copyright © 1972, John Wiley & Sons, Inc. Reprinted by permission of John Wiley & Sons, Inc.

Polytoluene: All of the absorptions of polystyrene present plus a new absorption at 1700 cm^{-1} as observed in polystyrene.

Polybenzene: All of the absorptions of polystyrene present plus a new absorption at 1720–1650 cm^{-1}, as observed in polystyrene.

Poly-p-Xylene: All of the absorptions of polystyrene present plus new absorptions at 2230–2175 cm^{-1} (disubstituted C$\equiv$C stretching vibration) and 1670–1600 cm^{-1} (CH out-of-plane deformation in RRC$=$CHR).

Polyhexane: The following absorptions were present: 3000–2900 cm^{-1} (H stretching vibration in CH$_3$, CH$_2$, and CH groups); 2250–2150 cm^{-1} (disubstituted C$\equiv$C stretching vibration); 1650 cm^{-1} (terminal C—C stretching vibration or internal cis CH$=$CH); 1470 and 1385 cm^{-1} (CH deformation in C—CH$_3$).

Polyisoprene: The following absorptions were present: 3030–2900 cm^{-1} (CH stretching vibration in CH$_3$, CH$_2$, and CH groups); 2210–1270 cm^{-1} (disubstituted C$\equiv$C stretching vibration); 1670–1610 (terminal C$\equiv$C or internal CH$=$CH stretching vibration); 1450–1375 cm^{-1} (CH deformation in C—CH$_3$).

Poly(vinyl chloride): All of the absorptions of PVC except for absorption at 700 cm^{-1} are present. Two new absorptions at 800 cm^{-1} and 750 cm^{-1}, both of which are present in a rubber hydrochloride.

Polytetrabutyltin: 3000–2800 cm^{-1} (CH stretching vibration in CH$_3$, CH$_2$, and CH groups); 1600 cm^{-1} (absent in monomer; conjugated C$=$C or C$=$O stretching vibration); 1450 cm^{-1} and 1385 cm^{-1} (CH deformation in C—CH$_3$).

The most common particle-size diameter for polystyrene was observed from electron micrographs to be 2357 Å. The particle size distribution, as illustrated for polystyrene in figure 6.10, shows clearly the uniformity of sphere diameter in the range ∼ 2140–2500 Å (also see figure 6.11).

Bell and his coworkers [25] obtained powder particles ranging in size from 0.2 to 1.2 μm also in a r.f. plasma range. They used a Bell-jar type of reactor, however, that contained two copper disc electrodes placed 5 cm apart; ethylene monomer was introduced at various rates in the absence of inert gas. They observed that operation at low flow and pressures produced powder, irrespective of the type of substrate used. The powder particles were highly crosslinked, however, and disappeared rapidly on the surface when the flow rate was increased. The material was insoluble in boiling m-Xylene.

Acetylene forms only a powder under all conditions used with the above-mentioned equipment [25]. Vinyl chloride produced only a powder at higher flow rates. Vinyl fluoride is similar to vinyl chloride except that the rate of powder formation is markedly reduced at higher flow rates. The infrared spectra of plasma powder obtained under the aforementioned conditions are

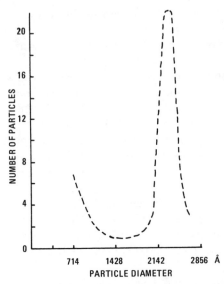

Figure 6.10. Particle-size distribution for r.f. plasma-polymerized styrene powder. *Source:* Adapted from R. Liepins and K. Sakaoku, *J. Appl. Polym. Sci., 16,* 2633 (1972). Copyright © 1972, John Wiley & Sons, Inc. Reprinted by permission of John Wiley & Sons, Inc.

shown in figure 6.12 for vinyl chloride, vinyl fluoride, and acetylene. These three spectra show considerable similarity to each other.

The following features are noted:

The peak at 1600 cm^{-1}, C=C bonds, indicates a high degree of unsaturation in all three polymers.

Figure 6.11. Spheres of dispersed plasma-polymerized polystyrene powder. *Source:* R. Liepins and K. Sakaoku, *J. Appl. Polym. Sci., 16,* 2633 (1972). Copyright © 1972, John Wiley & Sons, Inc. Reprinted by permission of John Wiley & Sons, Inc.

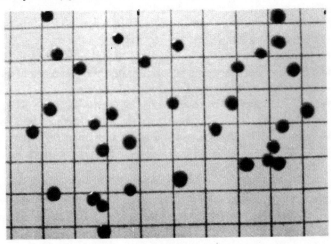

C=O bonds at 1700 cm^{-1} and OH bonds at 3400 cm^{-1} suggest that all three polymers are oxidized to some extent. This oxidation may have occurred upon removal from the discharge and exposure to air.

The peaks at 2960 cm^{-1} (CH$_2$ stretch) at 1463 cm^{-1} (CH$_2$ bending) and at 1369 cm^{-1} (CH$_3$ bending) underline the hydrocarbon nature of these polymers.

The absence of a doublet near 730 cm^{-1}, which is characteristic of crystalline hydrocarbons, reflects the amorphous structure.

The broad peak near 750 cm^{-1} (C—Cl stretch) for the vinyl fluoride polymer is the only characteristic that distinguishes these two polymers from the polymer deposited from acetylene.

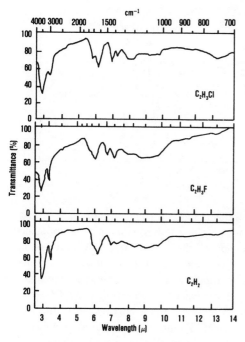

Figure 6.12. Infrared spectra of plasma-polymerized powders obtained from vinyl chloride, vinyl fluoride, and acetylene. *Source:* Adapted from H. Kobayashi, M. Shen, and A. T. Bell, *J. Macromol. Sci. Chem., A8*(8), 1345 (1974). Reprinted by courtesy of Marcel Dekker, Inc., N.Y.

From elemental analysis, the stoichiometry of the polymer deposited from vinyl chloride was found to be C$_2$H$_2$Cl$_{0.7}$, that from ethylene C$_2$H$_{2.7}$, and from acetylene C$_2$H$_{1.9}$ [25].

Recently, Auerbach et al. [26] investigated films obtained in a 3.9 MHz r.f., 17W glow discharge, tetramethyltin (TMT) and monobutyl trivinyltin (MBTVT) at a flow rate of 4.10–8.54 cm^3/sec. They observed that the stoichiometry of the starting compound is not retained in the deposited polymer. For instance, plasma films from MBTVT displayed a 15-fold change of the C/Sn ratio throughout the length of the reactor. It is noted that the film composition was influenced by the substrate, the C/Sn ratio being higher on glass depositions than on aluminum-deposited films.

Plasma polymers from silyl amines produced by a r.f. of 13.56 MHz were revealed [27] to be as hydrophobic as a plasma polymer from tetramethylsilane, which does not contain nitrogen. ATR-IR of polymers obtained from hexamethyldisilazane (HMDSZ), diethylaminotrimethylsilane (DATMS), and tetramethylsilane (TMS) are shown in figure 6.13.

Figure 6.13. ATR-IR spectra of plasma polymers from HMDSZ, DATMS, and TMS. *Source:* Adapted from T. Hirotsu, *J. Appl. Polym. Sci., 24,* 1957 (1979). Copyright © 1979, John Wiley & Sons, Inc. Reprinted by permission of John Wiley & Sons, Inc.

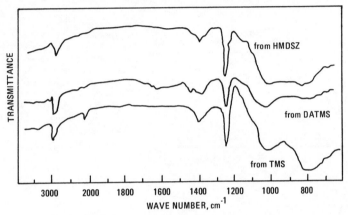

The broadening of these spectra as compared with those of the monomers suggests that the chemical structures of these polymers are more complicated. In the HMDSZ polymer the bands of the N—H stretching and bonding vibrations, observed at 3400 and 1550 cm^{-1} for the monomer, have almost disappeared. In contrast, Wrobel and colleagues [28] in studying the plasma polymerization of hexamethylcyclotrisilazane, reported that the IR spectra of the amino groups have essentially remained in the polymer. This difference may be due to the fact that the workers [28] employed a lower frequency of 20 KHz as compared with the 13.56 MHz used by Hirotsu. Another change in the spectrum is shown by a 1250 cm^{-1} band assigned to a trimethyl silyl group that has become much smaller in each polymer. The group may participate in the plasma polymerization, possibly by Si—Si coupling, which is frequently observed in silyl compounds.

It is also noted that a band at 2080 cm^{-1} has newly appeared in the tetramethylsilane plasma polymer. This band corresponds to the Si—H stretching vibration. The group may be formed by rearrangement of Si—CH$_3$. Another interesting feature is that all of the silyl plasma polymers exhibit the

Table 6.12. Elemental analysis of silyl plasma polymers estimated by ESCA peaks

Plasma polymer from	Element	Relative intensity of element		
		Monomer	Plasma polymer	Polymer/monomer
HMDSZ	C	6.0	6.0	1.0
	Si	2.0	5.76	2.88
	N	1.0	0.24	0.24
	O	—	0.24	—
DATMS	C	7.0	7.0	1.0
	Si	1.0	5.04	5.04
	N	1.0	0.35	0.35
	O	—	1.96	—
TMS	C	4.0	4.0	1.0
	Si	1.0	2.88	2.88
	O	—	0.17	—

Source: T. Hirotsu, *J. Appl. Polym. Sci., 24,* 1957 (1979). Copyright © 1979, John Wiley & Sons, Inc. Reprinted by permission of John Wiley & Sons, Inc.

Si—O—Si and/or Si—O—C bands that are detected around 1000–1100 cm^{-1} as the stretching vibration. These groups were evidently formed by the oxidation of radically activated species. The skeletons of these polymers suggest that they are highly crosslinked and composed of Si—C and Si—Si bonds, because the C—H group band around 2960 cm^{-1} has become much smaller. The authors [26] concluded from ESCA studies that the ease with which these silyl compounds fragment in plasma is on the order of N>C>Si. The elemental analysis of these polymers is given in table 6.12.

References

1. J. M. Tibbitt, M. Shen, and A. T. Bell. *J. Macromol. Sci., Chem., A10*(8), 1623 (1976).

2. P. L. Kronick, K. F. Jesch, and J. E. Bloor. *J. Polym. Sci., A-1*(7), 767 (1969).

3. H. Kobayashi, A. T. Bell, and M. Shen. *Macromolecules, 7,* 277 (1974).

4. H. Kobayashi, M. Shen, and A. T. Bell. Res. Contr., AD-762, 480, Office of Naval Res. (15 June 1973).

5. M. Seeger, R. J. Gritter, J. M. Tibbitt, M. Shen, and A. T. Bell. *J. Polym. Sci., Polym. Chem. Ed., 15,* 1403 (1977).

6. M. Seeger and E. M. Barrall II. *J. Polym. Sci., Polym. Chem. Ed., 13,* 1515 (1975).

7. J. M. Tibbitt, A. T. Bell, and M. Shen. *J. Macromol. Sci., Chem., A10,* 1617 (1976).

8. K. Jesch, J. E. Bloor, and P. L. Kronick. *J. Polym. Sci., A-1,* 4, 1487 (1966).

9. W. W. Knickmeyer, B. W. Peace, and K. G. Mayhan. *J. Appl. Polym. Sci., 18,* 301 (1974).

10. H. Kobayashi, A. T. Bell, and M. Shen. Proj. No. NR 256–526, Office of Naval Res. (1 Dec. 1973).

11. H. Schuler and L. Reinebeck. *Z. Naturforsch.*, *9a*, 350 (1954).

12. H. Schuler and L. Reinebeck. *Naturwiss.*, *19*, 445 (1952).

13. H. Yasuda and T. Hsu. *J. Polym. Sci., Polym. Chem. Ed.*, *15*, 2411 (1977).

14. H. Yasuda and T. S. Hsu. *J. Polym. Sci., Polym. Chem. Ed.*, *16*, 415 (1978).

15. A. R. Westwood. *Europ. Polym. J.*, *7*, 377 (1971).

16. D. T. Clark and D. Shuttleworth. *J. Polym. Sci., Polym. Chem. Ed.*, *17*, 1317 (1979).

17. L. F. Thompson and K. G. Mayhan. *J. Appl. Polym. Sci.*, *16*, 2291 (1972).

18. M. M. Millard, J. J. Windle, and A. E. Pavlath. *J. Appl. Polym. Sci.*, *17*, 2501 (1973).

19. M. Niinomi, H. Kobayashi, A. T. Bell, and M. Shen. *J. Appl. Polym. Sci.*, *18*, 2199 (1974).

20. L. F. Thompson. Ph. D. Dissertation, University of Missouri, Rolla, Mo. (1970).

21. M. R. Havens, K. G. Mayhan, and W. J. James. *J. Appl. Polym. Sci.*, *22*, 2793 (1978).

22. M. R. Havens, K. G. Mayhan, and W. J. James. *J. Appl. Polym. Sci.*, *22*, 2799 (1978).

23. L. F. Thompson and K. G. Mayhan. *J. Appl. Polym. Sci.*, *16*, 2317 (1972).

24. R. Liepins and K. Sakaoku. *J. Appl. Polym. Sci.*, *16*, 2633 (1972).

25. H. Kobayashi, M. Shen, and A. T. Bell. *J. Macromol. Sci., Chem.*, *A8*(8), 1345 (1974).

26. R. A. Auerbach, E. Kny, L. Leonard, and W. J. James. *Thin Solid Films, 64*, 395 (1979).

27. T. Hirotsu, *J. Appl. Polym. Sci.*, *24*, 1957 (1979).

28. A. M. Wrobel, M. Kryszowski, and M. Gaziecki. *Polymer, 17*, 678 (1976).

7. Properties of Plasma Polymers

1. Wettability

In surface wettability the effect of polarity is greater than that of the London forces (dispersion forces). The energy of a single dispersion force bond is approximately one Kcal/mole. On the other hand, a molecule physically adsorbed to a surface at, say, 50 segment sites, can be bonded via London forces with a total energy of about 50 Kcal/mole. This total bond energy due to dispersion forces has the magnitude of a chemical bond. It has been shown [1] that the (London) dispersion forces can be measured separately from the Keesom polar contributions to the total surface interaction. Since in surface wetting the effect of polarity outweighs that of the dispersion forces, any interference on the surface of polarity by even a very thin layer of nonpolar material will adversely affect wetting. Wettability, commonly defined by the contact angle between a drop of a liquid and the surface, is known to be an important characteristic that relates to the adherence of dyes, inks, and adhesives to a material. A small contact angle indicates that the liquid is wetting the material surface effectively; large angles show that wetting is poor.

Many methods have been used to increase the polymer surface energy, that is, improve wetting. Those methods include chemical, corona, flame, and plasma treatment. Table 7.1 shows the effect of various surface preparation techniques in altering the contact angle between different polymers and distilled water.

It is evident from table 7.1 that plasma treatment is generally more effective in improving wettability than any of the other techniques listed. Contact angles, as a measure of wettability, are generally obtained using a contact angle goniometer with a telescope and a stage mounted on an optical bench. Since trace contaminants on the surface may greatly affect the results, the polymer is usually soaked in a solvent, followed by washing with soap, rinsing with deionized water, and drying in a vacuum oven. Plasma-treated specimens should be allowed to stand several hours prior to measurement.

Table 7.1. Effect of surface treatment on wetting

	Contact angle (degrees)				
Polymer	Control	Plasma	Flame	Corona	Chemical
Polypropylene	87	22	87	35	60
Polyester	71	18	—	—	75
PVC	90	35	—	—	79
Polycarbonate	75	33	—	—	76
Silicone rubber	93	17	—	—	—
Polyethylene, HD	87	42	38	42	54

Source: Author.

It has generally been found that the contact angle of deionized water on a plasma-treated surface decreases with the time of exposure. Thus, for high- and low-density polyethylene, the contact angles decreased with treatment time in helium plasma, as shown in table 7.2. These results were obtained at 50 W power, 0.4 mm Hg and 50 cm^3/min He.

Table 7.2. Polyethylene contact angles versus deionized water (°C)

	Untreated	He (5 sec)	He (8 sec)	He (30 sec)
High density	96.5	85	84	81
Low density	97	76	87	68

Source: C. A. L. Westerdahl, J. R. Hall, D. W. Levi, and M. J. Bodnar, Tech. Rept. 4279, Picatinny Arsenal, Dover, N.J. (1972).

Frequently, contact angle measurements using a variety of test liquids are analyzed by a two-parameter surface energy model [2] to isolate the (London, d) dispersion, γ_{sv}^d, and (Keesom, p) polar, γ_{sv}^p, contributions to the solid vapor surface tension, γ_{sv}.

$$\gamma_{sv} = \gamma_{sv}^d + \gamma_{sv}^p \qquad \text{(Eq. 7.1)}$$

The technique of analyzing contact angle data (W_a) according to the above surface energy model has been detailed by Kaelble [3] and is summarized in equations 7.2–7.6.

$$\gamma_{LV} = \gamma_{LV}^d + \gamma_{LV}^p = \alpha_L^2 + \beta_L^2 \qquad \text{(Eq. 7.2)}$$

$$\gamma_{SV} = \gamma_{SV}^d + \gamma_{SV}^p = \alpha_S^2 + \beta_S^2 \qquad \text{(Eq. 7.3)}$$

$$W_a = \gamma_{LV}(1 + \cos\theta) \leq 2\gamma_{LV} \qquad \text{(Eq. 7.4)}$$

$$W_a = 2[\alpha_L\alpha_S + \beta_L\beta_S] = W_a^d + W_a^p \qquad \text{(Eq. 7.5)}$$

$$W_a/2\alpha_L = \alpha_s + \beta_s[\beta_L/\alpha_L] \qquad \text{(Eq. 7.6)}$$

where

γ_{LV} = liquid-vapor surface tension

γ_{SV} = solid-vapor surface tension

α_L, β_L = square root of the respective (London) dispersion, γ_{LV}^d, and (Keesom) polar, γ_{LV}^p, parts of γ_{LV}.

α_S, β_S = square root of the respective dispersion, γ_{SV}^d, and polar, γ_{SV}^p, parts of γ_{SV}.

W_a = nominal work of adhesion.

θ = liquid-solid contact angle.

In figure 7.1 the surface-energy components for various thicknesses of plasma-polymerized polyacrylonitrile, PPAN, on Teflon and on etched aluminum are illustrated. It is noted that on both substrates the value of the dispersion surface-energy component, γ_{SV}^d, is left almost unchanged.

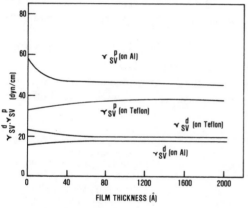

Figure 7.1. Variation of the surface energy components with film thickness for plasma-polymerized PPAN on Teflon and etched Al 2024-T3. *Source:* Adapted from C. L. Hammermesh and P. J. Dynes, *J. Polym. Sci., Polymer Letters Ed., 13,* 663 (1975). Copyright © 1975, John Wiley & Sons, Inc. Reprinted by permission of John Wiley & Sons, Inc.

On the other hand, the polar surface-energy component, γ_{SV}^p, is found to have altered considerably. On Teflon, a very nonpolar polymer with $\gamma_{SV}^p \simeq 1$ dyn/cm, conversion to a much more hydrophylic surface with $\gamma_{SV}^p \simeq 38$ dyn/cm is accomplished by PPAN films as thin as a few hundred angstroms. In contrast, deposition of PPAN on fresh FPL-etched aluminum causes a decrease in γ_{SV}^p from the very high value of ~ 58 dyn/cm to ~ 45 dyn/cm.

It was observed [1] that the initially high polar surface energy of PPAN films decreases somewhat with exposure to air. Within about 10 hours the value of γ_{SV}^p drops from ~ 45 to ~ 31 dyn/cm without significant changes in the dispersion surface energy component, γ_{SV}^d (see figure 7.2).

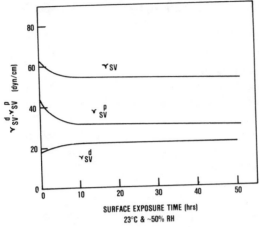

Figure 7.2. Variation in the surface energy components for a 1700 Å film of PPAN on etched aluminum exposed to air at 23°C and ~50% R. H. *Source:* Adapted from C. L. Hammermesh and P. J. Dynes, *J. Polym. Sci., Polymer Letters Ed., 13,* 663 (1975). Copyright © 1975, John Wiley & Sons, Inc. Reprinted by permission of John Wiley & Sons, Inc.

The observation of the changing character of a fresh plasma-deposited polymer has frequently been made and is related to a reaction between the residual and trapped free radicals and air oxygen. This is evidenced by the appearance of $-C{=}O$ and $-OH$ groups in the IR spectrum of plasma films exposed to air.

An interesting experiment was recently conducted [1] by measuring the effect on surface wettability of applying alternating layers of polar and nonpolar polymer on Teflon (figure 7.3). It is evident that the polarity of the uppermost layer has the greatest influence. The somewhat reduced effect of

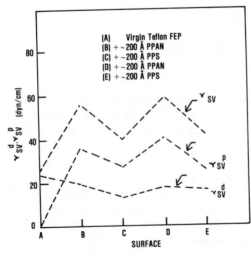

Figure 7.3. Effect of alternating ~200 Å layers of plasma-polymerized polyacrylonitrile, PPAN, and polystyrene, PPS, on the surface-energy components of Teflon FEP. *Source:* Adapted from C. L. Hammermesh and P. J. Dynes, *J. Polym. Sci., Polymer Letters Ed., 13,* 663 (1975). Copyright © 1975, John Wiley & Sons, Inc. Reprinted by permission of John Wiley & Sons, Inc.

Table 7.3. Effect of constitution of friction and wettability of halogenated polyethylenes

Polymer	Static coefficient of friction	Critical surface tension (dynes/cm)
Polyvinylidene chloride	0.90	40
Polyvinyl chloride	0.50	39
Polyethylene	0.33	31
Polyvinyl fluoride	0.30	28
Polyvinylidene fluoride	0.30	25
Polytrifluoroethylene	—	22
Polytetrafluoroethylene (Teflon)	0.04	18

Source: W. A. Zisman, "Surface Properties of Plastics," *Record of Chem. Progress*, 26(1), 23 (1965).

the PPS layer may be due to the presence of unreacted acrylonitrile on the surface, followed by incorporation into the forming polystyrene layer.

Often, the quantity critical surface tension, γ_c, is employed to characterize the solid surface. A semiempirical technique developed by Zisman [4] has been used to determine γ_c of the solid (table 7.3).

Washo [5] recently estimated the structure of a fluoron carbon surface after various degrees of plasma degradation from critical surface-tension data (table 7.4).

As a consequence of such surface degradation, crosslinking takes place, and unsaturated bonds may be formed. As early as 1966, Hansen and Schonhorn [6] observed that exposure of polyethylene surfaces to helium plasma for various times does not measurably change its surface tension of ~ 35 dynes/cm. Rather, the surface becomes highly crosslinked. The authors reported that other gases, such as argon, krypton, neon, and xenon, and even hydrogen and nitrogen, were all effective crosslinking agents, and only nitro-

Table 7.4. Structure versus critical surface tension, γc

Surface structure	γc
$-CF_2-$	~18
$>C-$ with F below	~34
$>C<$	~50

Source: B. D. Washo, IBM Corp., Poughkeepsie, N.Y. (1977).

gen caused a change in the wettability of the polymer's surface. The formation of free radicals during this process had already been reported in 1961 by Wall and Ingalls [7]. They concluded from solubility experiments that crosslinking occurred.

2. Adhesion

A. Introduction

It has generally been observed that exposure of many polymers and some metals to an activated gas plasma results in enhanced bond strength when these materials are subsequently bonded by adhesives. This effect may be due to several causes. Thus, crosslinking in the polymer surface results in an increased cohesive strength in the boundary layer and therefore enhances its resistance to cohesive failure. In other cases, dipoles are introduced into the surface in a plasma, such as by oxidation or grafting to enhance surface adhesion. It has also become recognized that removal by plasma treatment of ultrathin films of contaminants results in marked increases in bond strength. More recently, workers at MIT confirmed some theories voiced in the fifties that zeta potential and adhesion are interrelated [8]. They demonstrated that both zeta potential and peel strength of polyethylene increase nearly fourfold when the polymer is exposed to a helium glow discharge. Wettability of a material surface, an important condition for high adhesion, is also known to be greatly increased by a plasma treatment. Finally, etching of a polymer or metal surface in a plasma has been used to enhance adhesive bonding.

The mechanistic aspects of plasma contribution to adhesive bonding are not fully understood. As an example, in comparing the effect of oxygen plasma with that of helium on the adhesion of a number of polymers, workers have made the following observations [9].

High-density polyethylene responds about equally to helium or oxygen.
Polystyrene responds more to helium than to oxygen.
Polycarbonate responds more to oxygen than to helium.
Polypropylene showed more than eight times the original bond strength after oxygen plasma treatment for 30 min, while helium treatment showed no improvement over the control.

Some of these results can be explained in terms of a weak surface boundary layer that is reinforced by extensive crosslinking in a glow discharge. This surface treatment technique has frequently been referred to as CASING— Crosslinking by Activated Species of Inert Gases [6].

B. CASING

Hansen and Schonhorn [6] suggested earlier that while extensive interfacial contact between the adhesive and the polymer surface is necessary for the attainment of highest bond strength, this condition is not sufficient for forming strong joints. For example, when untreated polyethylene films are bonded at temperatures below their melting range, only weak adhesive bonds are obtained, even with highly wetting adhesives. These investigators suggested that such poor joint strength results from cohesive failure in the weak boundary layer caused by the presence of low-molecular weight polymer at the surface. They suggest that the primary function of surface oxidation techniques is to remove the low-molecular-weight weak boundary layer, and they point out that if surface oxidation or introduction of other polarizable groups alone occurred, without removal of the weak boundary layer, only weak adhesive joints would be obtained.

The authors [6] have reported that low cohesive strength of the weak boundary layer, which prevents the formation of strong adhesive joints, can be increased rapidly by permitting electronically excited species of rare gases to impinge upon the surface of polymers. If, for instance, these metastable and ionic gases come in contact with a polymer such as polyethylene, they cause abstraction of hydrogen atoms and formation of polymer radicals at and near the surface of the polymer. These radicals will interact to form crosslinks and unsaturated groups without scission of polymer chain links. Thus, cohesive strength is increased in the surface region by the formation of a dense gel matrix. Wettability of the surface is not measurably affected by these processes unless air oxidation is permitted to occur. As an example, the critical surface tension of polyethylene was found to be approximately 35 dynes/cm before and after plasma treatment.

Hansen and Schonhorn [6] referred to this surface treatment technique as "CASING." The effect of such CASING is illustrated in figure 7.4, which shows the effect of inert gas treatment of polymer samples in an electrodeless r.f. discharge (~100 W) on adhesive bond strength. The polymer film was sandwiched between aluminum panels using an epoxy resin as adhesive.

It was observed that contact time of activated gas with polymer film of as little as 1 sec resulted in markedly improved adhesive joint strength for polyethylene. Longer contact times were found to be required for polymers such as polytetrafluoroethylene. Helium, argon, krypton, neon, xenon, and hydrogen proved to be effective in the enhancement of adhesive strength in these experiments without measurably altering the wettability of the surfaces. Nitrogen, which also caused an increase in the bond strength, effected an increase in wettability, presumably by introduction of nitrogenous groups in the polymer surfaces.

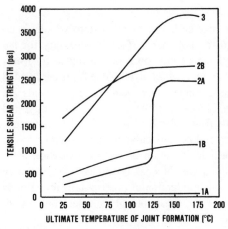

Figure 7.4. Effect of inert-gas plasma on the tensile shear strength of lap shear composites: (*1A*) untreated Teflon film, (*1B*) Teflon film + 10 min CASING, (*2A*) untreated polyethylene (Marlex 5003) film, (*2B*) Marlex 5003 + 10 sec CASING, (*3*) aluminum-epoxy resin-aluminum. *Source:* Adapted from R. H. Hansen and H. Schonhorn, *J. Polym. Sci., Polymer Letters Ed., 4,* 203 (1966). Copyright © 1966, John Wiley & Sons, Inc. Reprinted by permission of John Wiley & Sons, Inc.

It is noted that while both crosslinking and chain scission result from high-energy electron bombardment [10,11,12] of polymer surfaces, the CASING technique was found to produce only crosslinking and no measurable chain scission. Mass spectrographic analysis of the effluent gas during CASING of polyethylene showed that only hydrogen was generated; there was no evidence of gaseous hydrocarbons and no significant weight loss after 6 hr of bombardment with activated helium. The polyethylene surface became insoluble in trichlorobenzene at 135°C.

Thus, it can be seen from figure 7.4 that a tenfold or greater increase in adhesive joint strength was produced by such helium plasma treatment, although no change in wettability of the polymer surface occurred. The sharp rise in bond strength for the untreated polyethylene lap shear composites probably results from a melting of the polymer onto the cured epoxy adhesive. Infrared examination of the treated polyethylene film by attenuated total reflectance techniques disclosed only the formation of transethylenic unsaturation at the surface. Transmission spectra of treated and untreated films were identical, indicating that unsaturation occurs only at or near the surface during such glow discharge [6]. No carbonyl or hydroxy groups were detected.

It is assumed that such inert-gas plasma treatment also induces crosslinking in polytetrafluoroethylene. Thermograms of perfluorinated hydrocarbons disclosed that exposure to inert-gas plasma results in polymers of greatly increased molecular weight [13]. Transethylenic unsaturation in treated polytetrafluoroethylene surfaces was also discovered by infrared techniques. When those surfaces were brominated, no increase of adhesive joint strength was observed as compared with the treated but unbrominated specimens, indicating that such unsaturation does not play a significant role in the formation of strong adhesive joints. Therefore, Schonhorn and Hansen [13] concluded from their experiments that the observed considerable improvement in adhesive joint strength by such CASING technique is chiefly due to the formation of a crosslinked surface layer that has high cohesive strength.

It is noted that such surface reactions occur much more rapidly in amorphous than in crystalline materials. As an example, n-hexatricontane ($C_{36}H_{74}$), which is highly crystalline and melts at 70°C, remained essentially unchanged when exposed to helium plasma at 30°C but showed a greatly increased molecular weight when exposed to the same plasma conditions in its molten state. The investigators [13] suggested, therefore, that in the surfaces of partially crystalline polymers, only the noncrystalline regions are crosslinked.

The thickness of the crosslinked layer, as determined by gel fraction analysis, was found to be related to the time of plasma exposure [13]. For polyethylene in helium plasma, the estimated thickness increased from 30 Å at 1 sec exposure to ~10,000 Å for about 1000 sec exposure. Figure 7.5 illustrates the dependence of tensile shear strength on exposure time [13]. The results show that maximum tensile shear strength is obtained after about 5 sec exposure to helium plasma. This corresponded to a crosslinked surface thickness of 200–500 Å. This appears to be the upper limit of the weak boundary layer in the untreated polyethylene.

It is noted that the wettability requirements for the epoxy polyethylene system were satisfied ($\theta \simeq 0$). Therefore, Schonhorn concluded that in melt-crystallized films of polyethylene, a weak boundary layer was generated at the liquid-air interface. This conclusion was supported by showing that strong adhesive joints are formed in a poly(chlorotrifluoroethylene)-epoxy system using a conventional epoxy adhesive ($\gamma_{LV} \simeq 33$ dyne/cm) [14]. Schonhorn [13] suggested therefore that the polyhalocarbon does not have an inherently

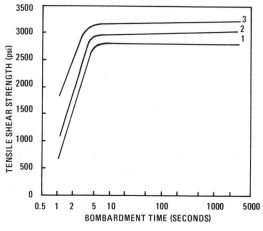

Figure 7.5. Tensile shear strength of the composite aluminum-epoxy adhesive-He CASED polyethylene-epoxy adhesive-aluminum versus exposure time (1 mm Hg He) at (1) 60°C, (2) 82°C, and (3) 104°C. *Source:* Adapted from H. Schonhorn and R. H. Hansen, *J. Appl. Polym. Sci., 11,* 1461 (1967). Copyright © 1967, John Wiley & Sons, Inc. Reprinted by permission of John Wiley & Sons, Inc.

weak boundary layer, as opposed to polyethylene, which has the same critical surface tension of wetting [4].

It is noted that such CASING by plasma results in strong adhesive joints without changing the wettability of the polymer. These researchers suggest that other surface treatments, such as oxidation, are effective because they, like plasma CASING, remove the weak boundary layer normally present on the polyethylene surface. Surface treatments by methods such as corona discharge and chemical etching have also been shown [15] to produce crosslinking on polyethylene surfaces and strong joints. The mere presence of polar groups would not be effective in the buildup of strong bonds if these polar groups are located on a weak surface.

Polytetrafluoroethylene required the considerably longer plasma exposure of 1000–5000 sec as compared with polyethylene (5 sec) for maximum bond strength under comparable experimental conditions. Since the wettability of the polymer was found [13] to be unchanged and the surface tension of the adhesive relatively high with respect to achieving a reasonable degree of wetting, the results shown in figure 7.6 are surprising.

Figure 7.6. Tensile shear strength of composite aluminum-epoxy adhesive-Teflon-epoxy adhesive-aluminum versus exposure time in 1mm Hg neon plasma at (1) 60°C, (2) 82°C, (3) 104°C. *Source:* Adapted from H. S. Schonhorn and R. H. Hansen, *J. Appl. Polym. Sci., 11,* 1461 (1967). Copyright © 1967, John Wiley & Sons, Inc. Reprinted by permission of John Wiley & Sons, Inc.

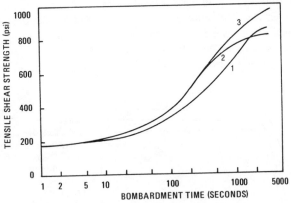

Poly(vinyl fluoride)-based composites showed an increase in tensile shear strength from ~500 psi to ~1800 psi when treated in a 1:1 mixture of H_2:He for 1 hr at 1 mm Hg pressure. On the other hand, exposure of poly(vinylidene fluoride) did not result in improved bond strength. This fact is taken [13] as an indication that this polymer, as normally prepared, does not have a weak boundary layer.

These results seem to be incompatible with the general association of polarity and the compatibility rule of de Bruyne [16]. It would appear that in addition to such CASING by plasma, free radicals formed in the polymer surface contribute to the observed enhancement of adhesive bond strength.

C. Composition of the Surface

Plasma is very similar to other ionizing radiation, such as γ-radiation, X-radiation, UV radiation, and high-energy electron beams, which can (1) initiate polymerization of certain monomers and create free radicals on polymers exposed that lead to (2) crosslinking of the polymer and/or (3) degradation of the polymer or (4) can be further utilized as the initiation sites of graft polymerization.

The characteristic features of plasma here are:

The radiation effect is limited to the surface and the depth of the layer affected is much smaller ($1-10~\mu$) than by other more penetrating radiation, and

The intensity at the surface is generally stronger than that by the more penetrating radiations [17].

The ultraviolet radiation in inert plasma will generate free radicals in many if not most polymer surfaces that interact with each other to form crosslinks in the outermost surface and therefore increase its molecular weight.

On the other hand, a mechanism for the photodegradation of polyethylene that involves oxygen in its excited state has been proposed [18]. Atomic oxygen, which has a heat of formation of 59.5 Kcal/mole, which is the order of organic covalent bonds, is known to attack polyethylene rapidly at room temperature. It has been reported that this leads to conversion of solid polyethylene into gaseous products. The rate of attack is reproducible and easy to control. Under milder plasma conditions, however, polymer degradation has been reduced and a compromise established in favor of introducing into a nonpolar surface a degree of polarity, such as carbon-oxygen or carbon-nitrogen moieties, which greatly enhanced adhesion.

The relationship between increased molecular weight of the polymer surface by crosslinking and adhesive strength is clearly related to the greater cohesive strength of the crosslinked material in the surface (see previous section). Noble gas plasmas tend to produce primarily molecular-weight increases (with little effect on wettability) because of crosslinking that penetrates below the surface to a depth of $1-10~\mu$, depending primarily upon the exposure time. The crosslinking mechanism is believed to be straightforward [13].

$$\text{RH} \xrightarrow[\text{activation}]{\text{Plasma}} \text{R} + \text{H}$$

$$\text{R}_1 + \text{R}_2 \xrightarrow{\hspace{3cm}} \text{R}_1 - \text{R}_2$$

Crosslinking densities can be large and equivalent to the density produced by a 100 MR dose of high energy radiation [19]. The degree of crosslinking in polyethylene is larger than that in polypropylene.

Molecular weight changes have been related indirectly by improved adhesion via the weak-boundary-layer concept [13]. Typical data are shown in figure 7.7 [20]. The crosslinked surface layer is insoluble in solvents. The gelation mass (crosslinked surface layer) can be measured and used to estimate the depth of crosslinking penetration. This mass grows with time, and the growth is proportional to the square root of time [13] (figure 7.8). This has been generally observed for all noble gas plasmas, hydrogen, and nitrogen plasmas [21].

Crosslinking theory is well defined and has been employed for quite some time to explain crosslinking caused by high energy radiation. The number of crosslinks at a given site per primary-weight average molecular weight has been mathematically related to the unit wavelength [19] and the experimentally determined gelation mass has been related to the theory [22]. It is interesting to note that early UV radiation experiments using high-pressure mercury UV lamps, which emit most radiation at 2537 Å, have not produced much crosslinking. Therefore, the effect of UV on crosslinking had been neglected in subsequent work. It has been recently observed, however, that the effective wavelengths for polyethylene crosslinking are below 1900 Å [23].

There are other parameters that affect molecular-weight changes upon exposure to a plasma. For instance, unsaturation can be produced simultaneously with the molecular-weight change. The presence of double bonds not only alters the mode of surface reactivity but also changes the UV absorption coefficient. The degree of crosslinking increases with plasma temperature. It is interesting to note that there is a large discontinuous change in the extent of crosslinking across the glass transition temperature [24]. This fact led to the suggestion that crosslinking occurs only in the amorphous regions of the polymer [13]. More recent crosslinking data with high- and low-density polyethylene, that is, containing high and low degrees of crystallinity, do not confirm this theory, however [25]. Impurities present in the polymer are also known to affect molecular-weight changes in plasma. They can change the UV absorption coefficient and can serve as active sites to initiate molecular-weight changes.

Figure 7.7. Bond strengths for polymers treated in oxygen and helium plasmas. *Source:* Adapted from J. R. Hall, C. A. L. Westerdahl, A. T. Devine, and M. J. Bodnar, *J. Appl. Polym. Sci., 13,* 2086 (1969). Copyright © 1969, John Wiley & Sons, Inc. Reprinted by permission of John Wiley & Sons, Inc.

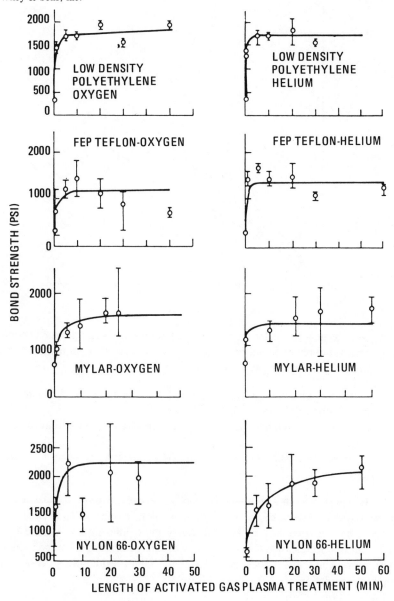

Figure 7.8. Effect of time on gelation. *Source:* Adapted from J. R. Hall, C. A. L. Westerdahl, A. T. Devine, and M. J. Bodnar, *J. Appl. Polym. Sci., 13,* 2086 (1969). Copyright © 1969, John Wiley & Sons, Inc. Reprinted by permission of John Wiley & Sons, Inc.

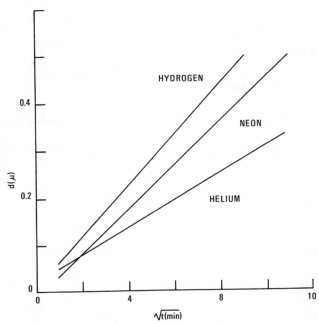

Atomic oxygen, for instance, as generated in plasma, is an important chemical species that reacts rapidly with organic substances at room temperature. This process is interesting because [26]:

The chemical changes on the resulting surface facilitate the formation of strong adhesive bonds.

Significant morphological features lying below the surface may be revealed.

Polymer can be cleanly removed from surfaces that are resistant to oxidation.

It has been shown [27] that the reactive species in oxygen plasma can convert surface layers of the polymer to volatile products. The remaining surfaces have low contact angles with water and exhibit improved adhesion. In fact, it has been shown [28] that a glow discharge in oxygen can be used to clean mirrors and other optical devices by the "rapid, residue-free, gentle and complete oxidation of contaminant films."

Introduction into the polymer surface of polarity such as carbon-oxygen moieties has been found to improve adhesion markedly in some polymers and less in others (table 7.5). Lap shear samples were bonded with Epon 828/ Versamid 140 epoxy system.

From these data it appears that not readily oxidizable polymers, such as polyether sulfone and polyaryl sulfone, do not show better adhesion by oxygen plasma than by helium plasma treatment. Since the polymer samples were exposed to helium plasma for 50 min as compared with 5 min exposure to oxygen, however, the data should be interpreted cautiously. The more surprising and perhaps less understandable results are those of Nylon 12, especially when compared with Nylon 11.

The incorporation of nitrogen into the polymer surface has been known to result in carbon-nitrogen bonds and has recently been measured by Electron Spectroscopy for Chemical Analysis (ESCA) [17]. From these data, a significant level of nitrogen incorporation is evident. Since nitrogen is one of the most electronegative atoms, a strong effect of such nitrogen incorporation into the polymer surface on adhesion might be expected.

Yasuda [17] concludes from ESCA data the existence of C-N bonds. He suggests that the nitrogen attachment to carbon represents the direct reaction of activated gases, while the oxygen incorporation by argon plasma may be due to the formation of free radicals in the polymer surface followed by free-radical reaction with oxygen. Whatever the mechanism, introduction of

Table 7.5. Effect of polarity on adhesion

Polymer	Surface treatment	Shear strength (psi)	Type of failure
Valox (aromatic polyester)	He plasma, 50 min	655	adhesive to steel
	oxygen plasma, 5 min	910	adhesive to steel
Polyether sulfone	methanol wipe, sand	615	adhesive to polymer
	He plasma, 50 min	730	adhesive to steel
	oxygen plasma, 5 min	580	adhesive to steel
Astrel (polyarylsulfone)	methanol wipe, sand	555	adhesive to polymer
	He plasma, 50 min	1040	adhesive to polymer
	oxygen plasma, 5 min	840	adhesive to polymer
Nylon 11	methanol wipe, sand	700	adhesive to polymer
	He plasma, 50 min	425	adhesive to steel
	oxygen plasma, 5 min	825	adhesive to steel
Nylon 6/6	methanol wipe, sand	425	adhesive to polymer
	He plasma, 50 min	575	adhesive to polymer
	oxygen plasma, 5 min	770	adhesive to steel
Nylon 12	methanol wipe, sand	830	adhesive to polymer
	He plasma, 50 min	900	adhesive to polymer
	oxygen plasma, 5 min	745	adhesive to steel

Source: M. C. Ross, U.S. Army, ARRADCOM, LCWSL, Tech. Rept., ARLCD-TR-77088 (1978).

Table 7.6. Electronegativities of various elements

Element	Electronegativity
Fluorine	4.0
Oxygen	3.5
Chlorine	3.0
Nitrogen	3.0
Bromine	2.8
Carbon	2.5

Source: Linus Pauling, *The Nature of the Chemical Bond,* 3d ed. Copyright © 1960 by Cornell University. Used by permission of the publisher, Cornell University Press. Cornell University Press: Ithaca, N.Y. (1960).

any polar atoms into the material surface is expected to enhance wetting and adhesion in less polar polymers. Electronegativities for some elements of common occurrence in organic molecules on the most widely used scale, that of Pauling [29], are listed in table 7.6. Thus, introduction of polar atoms such as fluorine, oxygen, nitrogen, chlorine, and so forth into the surface of a nonpolar material should be expected to enhance its wetting and adhesion.

Plasma oxidation studies of fluorocarbon polymers have demonstrated the existence of oxygen functional groups attached to the polymer and changes in wettability, adsorption, and adhesion [30]. Information on plasma oxidation of silicones has also been published [30,31]. In this case, the oxygen-containing functional groups are primarily hydroxyl and are attached to the methyl groups, that is, Si-CH$_2$OH. The reaction mechanism shown below has been suggested.

$$CH_3 - Si + O\cdot \rightarrow CH_2\cdot - Si + OH$$

$$C_2 - Si + OH\cdot \rightarrow CH_2OH - Si$$

Plasma oxidation of polymers containing nitrogen and/or oxygen is not confined to an oxygen plasma. Oxidation can be produced in a noble gas plasma as well. In these cases, the oxygen is supplied by the polymer, and the polymer free radicals are generated through bond dissociation reactions. Polymethylene [30] and polydimethylsiloxane [32] are two examples. Oxidation of polymer surfaces has been carried out in oxygen-containing gases, such as air, CO_2, N_2O, and so forth [31].

An oxygen plasma produces polymer free radicals that can crosslink, but subsequent oxidation prevents rebonding, resulting in a decrease in the molecular weight and formation of volatile compounds that ablate from the polymer surface. This process leaves a thin crosslinked layer of some 500 Å thickness below the surface, followed by a second layer of 1–10 Å thickness, which exhibits a decreased molecular weight [33].

Ablation produced a continuous and linear (with time) loss of material from the polymer surface. The rate of such ablation is also a function of the polymer composition, degree of branching, degree of crosslinking, and mechanical stress contained in the polymer. Rate of ablation decreases with polymer structure in the order indicated below.

| Hydrocarbon polymers containing oxygen | > | Hydrocarbon polymers | > | Fluorocarbon polymers |

Typical ablation data are shown in figure 7.9.

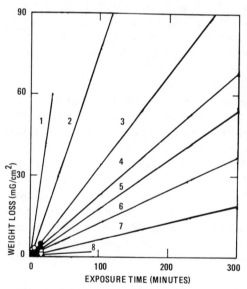

Figure 7.9. Ablation rate measurements for various polymers exposed to an oxygen plasma: (1) polysulfide, (2) poly(oxymethylene), (3) polypropylene, (4) low-density polyethylene, (5) poly(ethylene glycol terephthalate), (6) polystyrene, (7) polytetrafluoroethylene, and (8) sulphur-vulcanized natural rubber. *Source:* Adapted from R. H. Hansen, J. R. Hall, C. A. L. Westerdahl, A. T. Devine, and M. J. Bodnar, *J. Polym. Sci., 13,* 2085 (1969). Copyright © 1969, John Wiley & Sons, Inc. Reprinted by permission of John Wiley & Sons, Inc.

Amino groups have been attached to polymer surfaces by using ammonia or nitrogen and hydrogen plasma [34]. These groups are confined to a thin layer with dimensions that are less than 1 μ. In these reactions the polymer temperature was found to be an important parameter; increasing the temperature generally caused an increase in the thickness of the amino layer. It appears that a nitrogen-hydrogen gas mixture is more effective than ammonia. An interesting case of the use of amino groups introduced by plasma into a polyethylene surface is the subsequent treatment of the surface with an anticoagulant like heparin, which increases the clotting time of blood.

R. A. Auerbach has recently shown [35] in his work on organic species such as carbenes and nitrenes that difluorocarbene and trifluoromethyl radicals generated by plasma exposure from perfluorocyclobutane and hexafluoroethane, respectively, react with either organic or inorganic substrate surfaces to induce a high degree of hydrophobicity of Teflon-like properties—extremely poor wetting properties. Water contact angles on aluminum, stainless steel, polyethylene film, and polyester film exposed to perfluorocyclobutane plasma were found to be increased by 40%, −18%, and 130%, respectively. The small hydrophylic effect on polyethylene may not be significant in view of its hydrophobicity per se. In contrast, when metal and organic surfaces were exposed to hexafluoroethane in plasma, only the organic materials, such as polycarbonate, polyester, and polyethylene, displayed increased water contact angles, by 55%, 127%, and 18%, respectively.

Earlier work by Olsen and Osteraas [36] in nonplasma generation of difluorocarbenes, $:CF_2$, by pyrolysis of sodium difluoroacetate, indicates that a progressive modification of polymer surfaces by reaction with $:CF_2$ takes place. It was shown that this reaction is an irreversible grafting and imparts a high degree of hydrophobicity to polymer surfaces such as polyester, cotton, cellulose triacetate, rayon, and nylon, A recent U.S. patent [37] describes plasma grafting onto polymer surfaces of organic compounds, such as carboxylic acids, alcohols, esters, amines, nitriles, and vinyl derivatives. Claims included acids containing up to 10 carbon atoms; aliphatic mono and polyols and phenols having up to 10 carbon atoms, the esters derived from the aforementioned acids and alcohols; amines that were primary, secondary, or tertiary, aliphatic or aromatic, having up to 20 carbon atoms; aromatic and aliphatic nitriles having up to 10 carbon atoms; and unsaturated compounds of the vinyl, vinylene, or vinylidene groups attached to a carbonyl, sulfonyl, formyl, carboxy, halogen, or nitro group. The plasma conditions mentioned in this important and broad-claimed patent included (in addition to the aforementioned monomers in gas form) N_2O_3, water vapor, and noble gases under a r.f. field ranging from 0.5 to 30,000 MHz and a vacuum from 0.01 to 22 mm Hg.

Although there is a large volume of published data on grafting chemical entities to solid surfaces via high energy radiation, very little attention has been given to the use of plasma for this application. Most published work deals with the grafting of polymers onto the surface of fibers, metals, or other inorganic materials. Three methods of plasma grafting have been described. In one, the substrate is exposed to a discharge sustained in the gas or vapor of the monomer to be grafted. The weakness of this approach is that the monomer forms a polymer in the plasma without attaching itself to the desired surface or forming only a weak bond with the substrate.

Another potentially more successful approach is to expose only the substrate to the glow discharge. In this way, free radicals would be generated on the surface, which can be employed for grafting in two ways. In one the treated surface is contacted immediately after removal from the plasma with the monomer, either in the liquid or vapor phase. The surface free radicals then are available for chain initiation. This approach requires precautions to prevent a massive decay of the surface free radicals. By this method polymerization can start only at the surface of the plasma-treated material, and it should provide a better graft. Some homopolymerization may take place by this method because of a chain transfer mechanism, but the extent should be much less than by the first method.

A third approach is to expose the substrate to be treated to air or oxygen plasma in such a manner as to generate peroxides on the surface. Peroxides are known to serve as catalysts in free-radical polymerization. Such treated surfaces could be exposed to suitable monomers at a somewhat later time under the most effective conditions. The grafting in a corona discharge of a number of acrylates to cellulose was reported some time ago [38]. More recently, acrylonitrile was grafted to cotton fiber in a r.f. discharge of argon [39] with the formation of only negligible amounts of homopolymer. Di- and tetra-fluoroethylene were also reported to be successfully grafted onto cotton to obtain a hydrophobic surface.

Auerbach and O'Connor of Lord Corporation [40] disclosed in 1977 the treatment in argon plasma of a Kevlar (polyaramide) fabric followed quickly by immersion of the treated fabric in an epoxy resin. In this way the investigators obtained an increase of the cured fabric in interlaminar shear strength of approximately 45% over the untreated material. It appears that in this technique the free radicals that formed on the fibers reacted with the epoxy resin to form covalent bonds with the resin.

Grafting of monomers onto wool is a widely investigated method. Various chemical and radiation methods have been employed in the past to create active sites capable of initiating polymerization on the wool surface. Plasma grafting onto this material has been reported in more recent times [41,42]. Fluorocompounds were successfully grafted to wool in a plasma of argon and fluorocarbon monomers [43]. This is not practical in many cases, since many desirable monomers have no vapor pressure.

An important step was taken when Pavlath and Les [44] impregnated wool that had been plasma-treated just prior to impregnation with monomers dissolved either in water or in solvents. Since plasma pretreatment made the wool more hygroscopic, aqueous solutions displayed the most beneficial effect. The monomers employed in this study were tribromo-m-aminobenzene sulfonic acid in aqueous HCl solution and dibromopropyl diallylphosphate and borate in chloroform solution.

The effect of covalent grafting was determined by subjecting the treated fibers to the corresponding solvent. The graft was identified by X-ray photoelectron spectroscopy [42]. The amount of grafting was found to be on the order of 2% to 5% weight uptake.

D. Zeta Potential and Adhesion

All materials that are in contact with water or aqueous solutions display a potential difference against the so-called Grenzflächenpotential (interface potential), indicating that the material surface is electrically charged. This phenomenon is referred to as zeta potential.

The definition of the zeta potential is based on the concept of an electrical double layer at interface material/solution [45,46,47]. At the surface, which in pure water always displays a negative charge, a monomolecular layer of "Gegenionen" (antiions) adsorbs the so-called Stern layer [48]. Adjacent to it is the diffuse Gouy-Chapman layer [49,50], which in addition to co-ions also contains antiions needed for complete compensation of the material surface. The border line between the Stern layer and the Gouy-Chapman layer is referred to as the outer Helmholtz area, and the potential existing in this area is known as the Stern potential. The zeta potential is now defined as the potential gradient along the Gouy-Chapman layer and also from the outer Helmholtz area to the bulk phase of the liquid.

The zeta potential can be measured by various methods, such as the electrophoretic, electroosmotic, sedimentation-potential, and stream-potential methods. The last is the preferred method in the field of fiber characterization. By this method the zeta potential can be calculated using a formula by Helmholtz-Smoluchowsky:

$$\zeta = \frac{4 \pi \eta}{D} \cdot \frac{E}{p} \cdot \lambda(\nu) \qquad \text{(Eq. 7.7)}$$

where

ζ = zeta potential
η = viscosity of the liquid phase
p = pressure difference
D = dielectric constant
E = stream potential
λ = conductivity of the liquid phase.

It has been reported [51] that materials with a negative zeta potential inhibit the formation of blood clots when the material is inserted into the circulatory system, while positive zeta potential facilitates clot formation. It has also become known that the negative zeta potential of polar plastics can be increased by electrical polarization.

Stern's theory of the electrochemical double layer permits the determination of the adsorption potential of 1–1 valent electrolytes to solid interfaces. According to Stern [48] the following relationships apply:

$$\phi_- + \phi_+ = 2RT \ln c_{max} = Q$$
$$\phi_- + \phi_+ = 2F\zeta_{max}$$

where

ϕ_- and ϕ_+ = adsorption potential of anions and cations, respectively
c_{max} = electrolyte concentration at maximum ζ potential, ζ maximum
Q = molar heat of adsorption.

The zeta potential has been generally used as a measure of adsorptivity. For instance, it has been shown [52] that the magnitude of adsorptivity decreases with decreasing zeta potential. More recently, workers at MIT confirmed some theories voiced in the fifties that zeta potential and adhesion are interrelated [53] (see figure 7.10). They also demonstrated that the zeta potential of polyethylene increased when it was exposed to various plasma conditions (see table 7.7).

E. Effect of Corona Treatment on Adhesion

The search for methods to bond normally incompatible polymers has included the use of surface activation techniques that include plasma-discharge

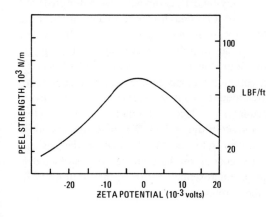

Figure 7.10. Peel strength versus zeta potential. *Source:* Adapted from N. Saka, G. Y. Yee, and N. P. Suh, presented in part at the Soc. of Plastics Engineers 35th Ann. Tech. Conf., Cleveland, April 25, 1977.

Table 7.7. Peel strength data for untreated and modified polyethylene

Bond specimen[a]	Peel strength (N/m)
Untreated polyethylene	2.77×10^2
Glow-discharge-treated polyethylene[b]	
Ethylene	4.23×10^2
Ethane	6.43×10^2
Carbon dioxide	7.25×10^2
Nitrogen	7.53×10^2
Ammonia	8.27×10^2
Helium	10.30×10^2
Doped polyethylene[c]	
LDPE/Aluminum	3.22×10^2
LDPE/Carbon black	5.32×10^2
Quasichemically treated polyethylene	
Witco Emcol E607	10.40×10^2
Ashland Variquat K-300	4.64×10^2

[a]Bond specimens formed at a pressure of 1.72×10^4 N/m^2.
[b]Treatment was at a frequency of 14×10^6 Hz, a pressure of about 1 mm Hg and a power of 100 watts for 1.0×10^3 seconds.
[c]The polyethylene was doped with 40% aluminum and carbon black by weight.
Source: N. Saka, G. Y. Yee, and N. P. Suh, presented in part at the Soc. of Plastics Engineers 35th Ann. Tech. Conf., Cleveland, April 25, 1977.

treatment of one or more of the components of a composite structure. Such plasma treatment may be conducted at low pressure, that is, in a glow discharge or near atmospheric pressure with a high current density (arc plasma) [54] or low current density (corona discharge). During or after plasma treatment, composites may be prepared directly by pressing together the activated polymer films [55] or by deposition of a second polymer on the activated surface [2,3,4].

Bond strengths of polypropylene-nylon 6/10 composites were reported [56] to be greatly improved by the use of a corona-discharge treatment. In these experiments, the nylon monomers were polymerized in situ on the treated polypropylene films. The weight losses of polymer were found to be negligible even after prolonged N_2 corona treatment. As an example, 0.01% weight loss was detected after 15 min corona treatment on one side of a 22 μ polypropylene film. For the short treatment times in these experiments, however, no weight changes were detectable. The effect of corona treatment on nylon peel strength and wettability is shown in figure 7.11. The investigators noted that a relatively smooth, continuous nylon coating formed on the corona-treated polymer surface, while the coating on the untreated film surface displayed a dropletlike layer. Optical microscopy confirmed that uniformity in coating was only obtained on the corona-treated surface.

Among the mechanisms proposed in interfacial adhesion, such as mechanical interlocking, adsorption, electrostatic interaction, and interdiffusion, the possibility of consolidation of a mechanically weak surface layer on the substrate has been mentioned [57], and its strengthening via CASING by glow discharge reported (see section B). Furthermore, certain types of surface structure may enhance capillary forces to improve the wetting of low-energy surfaces by a higher-energy liquid [58]. Finally, a partially "microporous" or locally permeable surface, namely, one with a large amorphous and microvoid content, would favor diffusion of adhesive or adhesive precursor molecules into the substrate, to produce an adhesive-substrate composition gradient instead of a sharply delineated contact interface. Such penetration at various sites is expected to enhance adhesion both by increasing the effective contact area and by reducing stress concentration. Chemical interaction between the surface and adhesive is not considered of primary inportance, since dispersion forces alone exceed the cohesive strength of most organic polymers [59]. The effect of corona treatment of polypropylene on the aforementioned increase in bond strength is not fully understood. The observed changes on the polymer surface include the formation of a surface layer of carbon-carbon unsaturation and crosslinking of 200–400 Å thickness. Nitrogen appears to behave as an inert gas, since nitrogenous groups were not detected in the corona-treated surface. Similar changes have been reported to occur by the glow discharge of polyolefin surfaces in helium (see the section 2B, "CASING").

It is interesting to note that although N_2 corona-treated polypropylene surfaces are oxygen-sensitive immediately after treatment, storage for 30 min under flowing nitrogen before exposure to air was found to inhibit subsequent oxidation [60]. As shown in figure 7.11, however, the surface energy of

Figure 7.11. Effect of corona duration on peel strength and polypropylene wettability. *Source:* Adapted from E. H. Cirlin and D. H. Kaelble, *J. Polym. Sci., Polym. Phys. Ed., 11*, 785 (1973). Copyright © 1973, John Wiley & Sons, Inc. Reprinted by permission of John Wiley & Sons, Inc.

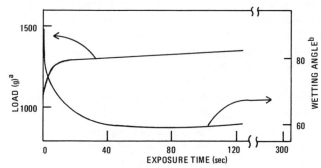

polypropylene was found to increase during N_2 corona treatment. Since similar changes in surface energy were observed with argon corona treatments, this change is apparently independent of the nature of the activated gas. It is not quite clear whether this effect is caused by the increased microscopic roughness of the surface or is the result of chemical modifications.

From a comparison of micrographs, N_2 corona treatment was estimated to produce a number of roughly hemispheral mounds rising to approximately 500 Å above the surface. The authors [56] suggest that these surface features may, at least in part, be the cause of the observed decrease in wetting angle, although some contribution to this effect would derive from the chemical surface modifications. In analogy to CASING [6] the consolidation of a weakly cohesive surface layer by corona-induced crosslinking may also be of importance in adhesion to polypropylene [61]. It has been suggested that the observed mounds are due to the presence of relatively disordered low-density regions, formed by the rapid thermal quenching that must follow fusion, in a matrix of unaltered polymer. The simultaneous crosslinking of the surface during corona treatment could restrict extensive reversion of the zones to the initial state. The authors [56] suggest also that the mounds may present regions of enhanced permeability through which nylon chains can grow or diffuse and ultimately interconnect the nylon coating to the polypropylene.

F. Contaminant Removal by Plasma

Recently, Jackson [62] reported on the effect of contaminants on adhesion and their removal by gas plasma, using for his tests circuit board panels printed with tinned copper on one side and epoxy laminate on the other. As representative types of contamination he employed conventional solder flux, photo resist, and vacuum pump oil, the contaminant levels ranging from concentrated (as received) to 0.0001 mg per 1.5 cm² surface area. Urethane and filled epoxy resin as normally used in encapsulation applications served as adhesives in these experiments. The effects of contaminant concentration on adhesion are summarized in table 7.8.

The results of this study indicated that both the type and concentration of the contaminant influences adhesion. Photo resist, that is, cured polymer, affects the adhesion of both the epoxy and the urethane more than solder flux (semisolid); vacuum oil (liquid) has the least effect. It appears that the maximum contaminant concentration that can be tolerated on printed circuit boards lies between 1 and 0.01 mg per 1.5 cm² surface area, while levels below 0.01 affect adhesion only moderately. Figure 7.12 illustrates the effectiveness of solder flux removal by argon plasma.

Contamination of a tinned-copper-printed circuit board with 1 mg of solder flux per 1.5 cm² surface area caused urethane adhesion failure. After exposure of the contaminated surface to argon plasma for 30–60 min, however, rela-

Table 7.8. Contaminant concentration level affecting adhesion of encapsulants

Contaminant concentration level	Solder flux	Vacuum oil	Photo resist
Concentrated			
1 mg/1.5cm^2	catastrophic failure (epoxy/diethanolamine)	catastrophic failure (epoxy/diethanolamine)	catastrophic failure (urethane)
1 to 0.01 mg/1.5cm^2	catastrophic failure (urethane)		catastrophic failure (epoxy/diethanolamine)
0.01 to 0.0001 mg/1.5cm^2	adhesion moderately affected (urethane)		adhesion moderately affected (urethane)
0.0001 mg/1.5cm^2		adhesion moderately affected (urethane)	

Source: L. C. Jackson, *Adhesives Age*, p. 34 (Sept. 1978).

Figure 7.12. Effectiveness of solder flux removal by argon plasma and urethane bond strength to printed circuit board surfaces. *Source:* Adapted from L. C. Jackson, *Adhes. Age,* p. 34 (Sept. 1978).

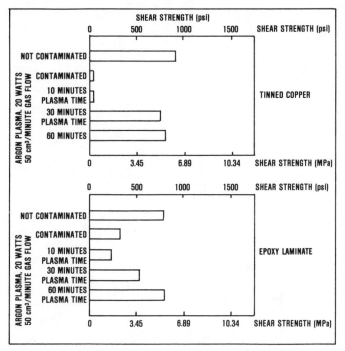

tively good adhesion was achieved. The epoxy laminate side of the printed circuit board required longer exposure times in plasma than the tinned copper side to remove the contaminant and establish satisfactory adhesion.

Balwanz [63] reported on the plasma cleaning of a large vacuum chamber, 55 × 12 feet, to remove water and other volatiles from its walls. He accomplished in an hour what would otherwise have taken months of heating and pumping. In one test case, beginning with walls contaminated with water and oil, 1 hr of nitrogen plasma cleaning made possible tank evacuation to 10^{-9} mm Hg. This rate of decontamination and degassing to such a high vacuum is considered unattainable by methods other than plasma treatment [63].

Recent attempts to clean metal surfaces reactively in a d.c. discharge at voltages ranging from 500 V d.c. to 3 kV d.c. were not successful [64]. On the other hand, r.f. discharges at 30 W and 0.5×10^{-2} mm Hg resulted in rapid cleanup of metal surfaces when an Ar-HCl plasma was used. In situ substrate surface analysis by SXAPS (X-ray Appearance Potential Spectrometry) indicated that reactive plasma cleaning in AR-HCl plasma at low powers is most effective in the removal of surface carbon and oxide layers.

The results as reported [64] indicate that such cleaning occurs by the formation of volatile compounds that are readily removed.

G. Adhesion to Metals

1. Adhesion of Plasma-Deposited Films to Metals

In considering the effect of plasma on adhesion, one generally distinguishes between the adhesion of plasma-treated surfaces in adhesive bonding and the inherent adhesion of plasma-deposited films. Data on the latter subject are sparse because of the lack of suitable techniques to obtain a true value of the adhesion of thin films to substrate surfaces. Dynes and Kaelble [65] have used a lap shear bond strength method to determine the cohesion and adhesion of plasma-polymerized films on aluminum. By this method, the apparent shear strength or cohesive strength of the film decreased with increasing thickness of the deposited film. It leveled off at about 1600 psi for films having a thickness of over 1600 Å. This method did not yield a quantitative value of adhesion, however.

James and colleagues [66] employed a "direct pull method" that provided a direct measure of the force required to pull the film from the substrate. At the same time this method showed whether the adhesive strength of the film was stronger or weaker than the cohesive strength of the film. Although this method also fails to provide values that can be translated into bond strength values, it nevertheless gives comparative indications of adhesive and/or cohesive strengths of plasma-deposited films on given substrates. The limitations of this method have been discussed in the literature [67]. The results of adhesion measurements of plasma-polymerized tetrafluoroethylene (PPTFE) and chlorotrifluoroethylene (PPCTFE) films on 304 stainless steel, aluminum, and silver are listed in table 7.9.

For PPTFE films on stainless steel, aluminum, and silver, the force per unit area is essentially the same; the mode of rupture was cohesive in nearly all

Table 7.9. Adhesion of plasma-polymerized films about 0.15 μm thick on various metals

Film	Metal substrate	F/A (kg/cm^{-2})
PPCTFE	304 stainless steel	94 $\pm$ 21
	aluminum	72
	silver	181
PPTFE	304 stainless steel	107
	aluminum	107
	silver	111

Source: K. Bhasin, D. B. Jones, S. Sinharoy and W. J. James, *Thin Solid Films, 45*, 195 (1977).

cases, that is, breakage took place within the film rather than at the interface. For PPCTFE, however, the strength of adhesion and/or cohesion was reported to vary with the substrate.

Recently Hammermesh and Crane [68] concluded on the basis of their experiments that adhesion of plasma-deposited films to polymers is a function of the chemical structure of the substrate and not of the plasma polymer. In addition, exposure of such bonds to various solvents, such as water, dimethyl formamide, hexane, and polyethylene glycol, for periods up to 120 hours, resulted in no deterioration of the bond strength of such films to the substrate polymers. It was, therefore, concluded that when adhesion occurs, a chemical bond is probably formed. On the other hand, these investigators have also shown [69] that for metals, unlike polymeric substrates, adhesion of plasma polymer is a function of both the nature of the substrate and the plasma-deposited film. Their data are shown in table 7.10.

The authors suggest that adhesion of polyacrylonitrile may be due to the more polar character of the plasma polymer as compared with polystyrene. Those polymers that adhered to metal were soaked in water at ambient temperature for 1 hr. Only the bonds of both plasma polymers to copper were found to be intact, because of the hydrophobic character of the copper, it was thought.

Using a large-volume microwave apparatus, other investigators [70] have observed that thiophene, when plasma-deposited onto carbon steel, displayed adhesion even after immersion at 270°C in a 10^{-4} lithium hydroxide solution at high pressure. More recently, similar results were obtained by Boenig and Blenner [71] by depositing thiophene in a r.f. discharge onto cold rolled steel. The bonds were found to exhibit excellent resistance to both water boiling for 5 hr and exposure to salt spray for 30 days.

2. Adhesion of Plasma-Treated Polymer Surfaces to Metals

In more recent times, considerable interest has centered on plasma treatment of polymer surfaces to enhance their adhesive bond strength to metal

Table 7.10. Adhesion of plasma films to metal substrates

| | Plasma-deposited | |
| | Polyacrylonitrile | Polystyrene |
Metal		
Aluminum	adheres	does not adhere
Steel	adheres	does not adhere
Nickel	adheres	adheres partially
Copper	adheres	adheres
Titanium	adheres partially	does not adhere

Source: C. L. Hammermesh and L. W. Crane, *J. Appl. Polym. Sci.*, 22, 2395 (1978). Copyright © 1978, John Wiley & Sons, Inc. Reprinted by permission of John Wiley & Sons, Inc.

Table 7.11. Effectiveness of plasma surface treatment of polymers on metal bonds

Polymer	Substrate	Test	Control[a]	Plasma treated
Polypropylene	steel	shear	12 psi	50 psi
ETFE[b]	copper	shear	8 psi	50 psi
FEP[c]	copper	shear	0 psi	50 psi
Polycarbonate	aluminum	shear	600 psi	1550 psi
TFE[d]	copper	180° peel	0 lb/in.	9 lb/in.
PVC	copper	180° peel	14 lb/in.	60 lb/in.
Polyethylene	steel needle	pull-shear	5.6 lbs	22 lbs

[a]No surface preparation.
[b]Fluoropolymer, Du Pont's Tefzel.
[c]Polytetrafluoroethylene-hexafluoropropylene copolymer, Du Pont's Teflon.
[d]Polytetrafluoroethylene, Du Pont's Teflon.
Source: L. Perrone, *Plastics Engineering,* p. 51 (1980).

surfaces. Perrone [72] pointed out that compared with other methods, such as chemical, corona, flame, or mechanical treatments, plasma offers lower operating costs. In fact, it had already been established during the sixties that exposure of many polymers to an activated gas plasma results in much stronger bonds when they are subsequently bonded by the use of adhesives, and more quantitative relationships have been studied during the seventies. As a result, for most polymers investigated, a plasma condition was found that produced greatly improved adhesive strength and frequently enhanced environmental bond resistance. The combination of features, such as low energy input, low-cost gases, and absence of polluting agents, in addition to enhanced bond quality, has already led to successful commercial application of this technology in the aerospace, electronics, and medical fields. Table 7.11 demonstrates the effectiveness of r.f. plasma treatment on some hard-to-bond polymers, such as some Teflons, a polyethylene, a polypropylene, and a polycarbonate on several metal substrates, when a general-purpose epoxy resin was used as adhesive.

It is evident from these data that treatment of the polymer surface by activated-gas plasma increased the adhesive bond strength in all cases by over 100% when compared with the untreated surface.

The relationship between wettability and plasma exposure of polymer surfaces has been discussed in section 1 of this chapter. Table 7.12 illustrates how increased wettability or decreased contact angle relates to tensile adhesion and plasma exposure.

Table 7.13 compares the effect of solvent cleaning of various polymers against exposure of the polymer to helium and oxygen glow discharge using Epon 828/Versamid 140 as the adhesive system (except in two cases where ethyl cellulose was employed). It is noted that the sulphur-containing polymers, polyethersulfone and polyphenylene sulfide, are not affected to any de-

Table 7.12. Effect of argon plasma treatment on contact angle and adhesive strength to metal

Material	Initial contact angle (°C)[a]	Initial tensile adhesion (psi)[b]	Plasma power level (watts)	Contact angle after plasma (°C)	New tensile adhesion (psi)[b]
Nylon 6/6[c]	48	2065	10	7	2807
Polymethyl methacrylate	54	513	10	40	1484
Nylon	37	2900	10	12	3440
Polystyrene	69	509	10	13	2050
Polyethylene	97	negligible	150	19	3780
RTV Silicone[d]	100	negligible	5	5	400

[a]After alcohol rinse and wipe.
[b]Materials bonded to aluminum tensile plugs with epoxy adhesive.
[c]The Polymer Corporation's Nylatron GS.
[d]General Electric's RTV 630.
Source: L. Perrone, *Plastics Engineering,* p. 51 (1980).

gree by such plasma treatment [73]. It is interesting to note the point of failure in these experiments. From the discussion in section 2B of this chapter, we may conclude that both polyarylsulfone and polyphenylene sulfide do not contain a weak boundary layer in the surface and are not, therefore, subject to a CASING mechanism.

In table 7.14 the effect of three different commercial surface treatments is compared with that of plasma exposure. It can be seen that except for polypropylene, plasma exposure results in superior bond strength when compared with other, more common surface treatments, such as chemicals, abrasion, flame, or corona. The advantage of chemical treatment of polypropylene over plasma exposure has been noted earlier in table 7.1 in connection with wetting properties [1]. The observations made in the first paragraph of section 2 seem to indicate that polypropylene requires exposure to oxygen plasma for extended times to realize maximum adhesive strength.

Plasma treatment in either argon or oxygen of molded rigid polyurethane foam was recently reported [69] to improve the bond strength to aluminum. Comparable bonding results were obtained with (13.56 MHz) r.f., (20 kHz) a.c., and d.c. plasma [74]. The treated surface retained improved bondability for extended periods of time. Both epoxy and RTV silicone adhesives were used in these experiments. The importance of these findings lies in the weaknesses experienced with the removal of mold-release agents. The plasma-treated parts exhibited foam failure in the bond test whenever optimum adhesive systems were employed.

Some exploratory work was recently reported in connection with vulcanized rubbers [75]. The surface was cleaned with hexane prior to helium plasma

Table 7.13. Lap shear bond strength of plastic-steel sandwich specimens

Plastic	Surface treatment time in minutes	Adhesive	Shear strength (psi)	Point of failure of adhesive material interface
Polybutylene terephthalate (Valox)	helium plasma, 50	A	665	steel
	oxygen plasma, 5	A	910	steel
	oxygen plasma, 10	B	2130	adhesive
Polyethersulfone (PES 200)	methanol wipe, sandpaper	A	615	polymer
	helium plasma, 50	A	730	steel
	oxygen plasma, 5	A	580	steel
Polyarylsulfone (Astrel)	methanol wipe, sandpaper	A	555	polymer
	helium plasma, 50	A	1040	polymer
	oxygen plasma, 5	A	840	polymer
Polyphenylene sulfide (Ryton)	methanol wipe, sandpaper	A	560	polymer
	helium plasma, 50	A	605	polymer
	oxygen plasma, 5	A	600	polymer
Ethylene-Chlorotrifluoroethylene Copolymer (Halar)	methanol wipe, sandpaper	A	370	polymer
	helium plasma, 50	A	820	steel
	oxygen plasma, 5	A	690	polymer
	helium plasma, 10	B	3860	polymer film failure
	oxygen plasma, 10	B	3830	polymer film failure
Nylon 11	methanol wipe, sandpaper	A	700	polymer
	helium plasma, 50	A	425	steel
	oxygen plasma, 5	A	825	steel
Nylon 6/12	methanol wipe, sandpaper	A	425	polymer
	helium plasma, 50	A	575	polymer
	oxygen plasma, 5	A	770	steel
Nylon 12	methanol wipe, sandpaper	A	830	polymer
	helium plasma, 50	A	900	polymer
	oxygen plasma, 5	A	745	steel

Source: M. C. Ross, U.S. Army, ARRADCOM, LCWSL, Tech. Rept. ARLCD-TR-77088 (1978).

Table 7.14. Plasma versus conventional surface treatments

Material	Surface treatment	Substrate	Load at failure (lbs)	Point of failure[a]
HDPE	plasma[b]	self	353.1	polymer jaw
	flame		144.1	bond
	corona		123.9	bond
	chemical		152.1	bond
Polypropylene	plasma[c]	self	110.4	bond
	plasma[d]		95.7	bond
	flame		30.7	bond
	corona		11.1	bond
	chemical		309.9	polymer lap
Polycarbonate	plasma[d]	self	154.3	bond
	abrasion		281.0	bond
	solvent		132.2	bond
Natural elastomer	plasma[d]	self	33.6	polymer lap
	abrasion		10.7	bond
	untreated		1.4	bond
PVC	plasma[d]	aluminum	5.0	polymer
	abrasion		2.20	polymer
	untreated		0.28	polymer
Polyester	plasma[d]	copper	4.41	polymer
	plasma	aluminum	317.0	aluminum
	abrasion		221.3	polymer aluminum[e]
	solvent		176.3	polymer
	chemical		11.6	polymer
	untreated		179.0	polymer aluminum[e]
Silicone elastomer	plasma[d]	aluminum	6.1	polymer
	abrasion		1.4	polymer
	solvent		1.1	aluminum

[a]Bond failure means interface between plastic surface and adhesive.
[b]Argon.
[c]Nitrous oxide.
[d]Oxygen.
[e]Failures evenly divided.
Source: L. Perrone, *Plastics Engineering,* p. 51 (1980).

exposure. Lap shear specimens were bonded to aluminum using Epon 828/ Versamid epoxy adhesive. Since the bonds were tested by peeling the specimens apart by hand, the results, shown in table 7.15, are considered to be qualitative only.

When plasma exposure times of polymers were plotted against bond strength to metals, the general shape of the curve appeared to be similar for different polymers [75]. A function was developed that describes the curves well enough to permit statistical examination. If the total strength of the adhesive bond after plasma treatment is S, and that of the untreated surface is S_0, the strength imparted by the activated plasma is

$$S = S_p + S_0 \qquad \text{(Eq. 7.8)}$$

The average strength contribution of the plasma per min treatment is obtained by dividing $S_p \ (=S - S_0)$ by the treatment time, t, in minutes.

$$R = \frac{S_p}{S_0} = \frac{S - S_0}{t} \qquad \text{(Eq. 7.9)}$$

where R is a function of time. A plot of R versus t suggests, from equation 7.9, that

$$\frac{1}{R} = a + bt \qquad \text{(Eq. 7.10)}$$

where a and b are constants, or

$$R = \frac{1}{a + bt}$$

Plots made using experimental points and plots calculated by equation 7.10 showed a good fit [75] except at very short times, at which the deviation was noticeable. The poor fit at very short times was ascribed to the fact that the $1/R$

Table 7.15. Effect of plasma treatment on the lap-shear bond strength of rubbers to aluminum

Treatment		Buna N[a]	Buna S[b]	Butyl[c]	Neoprene[d]
Untreated		peels easily	peels easily	peels easily	peels with difficulty in spots
Helium	30 min	not improved	improved a bit	improved a bit	improved a bit
NH_3	30 min	improved	not improved	not improved	good
Air	30 min	improved	improved a bit	improved a lot	good–best
O_2	30 min	improved a bit	not improved	improved a bit	improved a bit
N_2	30 min	improved	improved a bit	improved a lot	good

[a]Minor Rubber Co., Inc., Compound No. BA-7226, Durometer 60, 1/16″ sheet (MIL-R-3065B, SB610A₁B₁R₁F₁).
[b]Minor Rubber Co., Inc., Compound No. BR-14720, Shore A Durometer 60, 1/16″ sheet (MIL-R-3065B, RS625ABF₁).
[c]Minor Rubber Co., Inc., Grade CR-115, 1/16″ sheet.
[d]Minor Rubber Co., Inc., Compound No. BR-8098, Durometer 60, 1/16″ sheet (MIL-R-3065B, SC615A₁E₃F₁).
Source: C. A. L. Westerdahl, J. R. Hall, D. W. Levi, and M. J. Bodnar, Tech. Rept. 4279, Picatinny Arsenal, Dover, N.J. (Feb. 1972).

for very short times is not weighted heavily enough when the best straight line is drawn, since reciprocals of numbers of increasing magnitude become smaller. The investigators overcame the problem by using the least squares method to fit the best curve to the plot of R versus t. The equation for values for a and b was developed, and values for the linear correlation coeffecient $1/R = a + bt$ were listed [75].

The reader should remember that adhesion results, as well as, more broadly, all effects induced by glow discharges, should be taken with the understanding that the number of variables in a plasma experiment is considerable. It has been shown throughout this book that such variables include:

> type and concentration of positive and negative ions, free radicals, electrons, metastables, photons in an UN environment, power input, pressure, temperature, exposure time, monomer flow rate, plasma pretreatment and posttreatment, type and molecular weight of monomer species, nature of added gas (neon, argon, krypton, hydrogen, oxygen, nitrogen), presence and type of catalyst, concentration and type of monomer-gas systems, pretreatment and posttreatment of specimens, etching and ablation effects, chemical nature and energy of surfaces to be treated or plasma coated, inductive versus resistive coupling, continuous versus pulsed plasma, reactor design, character and fullness of glow, location in the glow region of specimens to be treated.

In addition, timing effects often appear to play an important role. In comparing the energies of the active species in plasma, ranging from 46 to 920 Kcal/mole, with the energies of covalent bonds in organic and inorganic molecules ranging from 37 Kcal/mole for the F-F bond to 143 Kcal/mole for the Si-F bond (with C-C and C-H bonds being 58.6 and 87.3 Kcal/mole, respectively) it would seem that under appropriate plasma conditions, all types of chemical bonds can undergo scission. The timed introduction of any of the aforementioned molecular species was found to be characteristic for the products formed in plasma as well as for the modification of the substrate surface in the reactor. It is unlikely, therefore, that all published data represent optimized or maximized results.

Hansen and colleagues [76] obtained a U.S. patent in 1969 that relates to the improvement of polymer adhesive bonding. Examples are cited to show high-tensile shear-strength values when hydrocarbons, fluorocarbons, and amide polymers are subjected to an inert-gas plasma prior to adhesive bonding to metals such as aluminum. The claim evaluates this technique with respect to improvement of bond strength between the plasma-treated polymer surface and an adhesive.

H. Adhesion of Plasma-Treated Polymer to Polymer

Very little has been published on the effect of plasma on the adhesion of plastic to plastic, plastic to rubber, or rubber to rubber. Enhanced bondability of many plasma-treated polymers to other surfaces has generally been related to increased polarity and thus wettability. Plasma-induced crosslinking strengthens the surface layer, thus reducing the potential failure due to a weak boundary layer, that is, cohesive strength [13].

It has long been known that for linear polymers, there exists a general relationship between mechanical properties, such as bending, impact, tensile, or abrasion resistance, and the molecular weight of the polymer. H. F. Mark showed as early as the thirties that up to a certain molecular weight, a material does not exhibit any significant degree of material properties. In terms of degree of polymerization, DP, the critical point, CP_c, at which linear polymers develop significant strength properties, lies for most species between 50 and 100. From this point on, these properties increase until into the DP range of 300 to 400, at which the strength–molecular weight curve converges to a point of diminishing gains. For tensile strength, TS, the relationship can be expressed in the form

$$TS = TS_\infty - C/DP_n \qquad \text{(Eq. 7.11)}$$

where

TS = tensile strength at the degree of polymerization DP_n
TS_∞ = tensile strength at infinitely high DP
C = a constant characteristic for a given polymer
DP_n = number average degree of polymerization.

Other factors adding to cohesion of polymers include, as defined by H. F. Mark, the supermolecular uniformity of polymer structure and texture, which includes the absence of stress risers or flaws.

Crosslinking by plasma causes the molecular weight of the polymer to increase in the surface up to a depth of about 10 μm. Since covalent bond distances range in the order of 1.5Å or 1.5×10^{-8} cm, a tighter structure is formed in the polymer surface, each additional crosslink adding a strength of about 2×10^{-4} to 5×10^{-4} dynes per bond. Furthermore, plasma-induced polarity may form hydrogen bonds in the bond interface, thus adding approximately 6×10^{-5} to 2×10^{-6} dynes per H bond.

In an interesting thesis published at MIT in 1977, Saka and colleagues [53] showed that the peel strength of polyethylene to itself via a butyl rubber adhesive is only marginally affected by the pressure, temperature, or duration of bonding. The peel strength was markedly increased by chemically modifying the polyethylene surface, however. It has also been argued [13] that the improvement in peel strength, when polyethylene is treated in a glow discharge, is due to elimination of the weak boundary layer on the surface by crosslinking. Furthermore, it has been stated [52] that the surface energy of polyethylene does not change as a result of crosslinking. This conclusion should be considered with some degree of caution, since evidence has been submitted in this book that free radicals are formed in the polymer surface, some of which have finite half-lives and therefore would alter the surface energies.

Saka and colleagues have also shown [53] that the zeta potential of polyethylene surfaces can be changed as a result of either chemical or plasma treatment and that the peel strength increases with increasing zeta potential until a maximum is reached. Since the zeta potential is an indirect measure of the surface charge, the investigators hypothesize that the relationship as shown in figure 7.10 explains what occurs on the polymer surface at the electronic level. They state that the peel strength is a bell-shaped function of the zeta potential value close to that of the adhesive (-9.6×10^{-3} V). They conclude that zeta potentials and the surface charges of the materials being bonded should be the same for good bonding and that this agrees with the observation that similar materials (or compatible materials) yield good bonding because their surfaces are electrically similar.

The same report showed that "doping" polyethylene with fillers such as carbon black resulted in a bond increase from 277 N/m to 535 N/m, with about a 10% increase in zeta potential. This agrees with data on a urethane system [77]. Goodyear polyurethane elastomer filled with acetylene carbon black showed a transition in zeta potential from negative to positive at approximately 28% of carbon black loading.

Chemical modification of a polyethylene surface, such as applying a boiling solution of quaternary ammonium salt in a 1:1 toluene-petroleum ether mixture followed by 1 hr exposure in an oven to 353°K, was also found to result in increases of both zeta potential and peel strength. Data for untreated and modified polyethylene, as well as for various plasma treatments, were shown in table 7.7.

It becomes evident from the preceding section that the mechanisms involved in the enhancement of bonding of plasma-treated polymer surfaces to metals via an adhesive actually relate to the enhancement of the adhesive quality between the polymer surface and the adhesive layer, regardless of the nature of substrate on the other side of the adhesive layer. Therefore, the data on adhesive bonding of polymer to metal are in this sense also applicable to polymer-to-polymer adhesive bonding.

Thus, the data given throughout this book and in many other published accounts indicate that the surface of most, if not all, polymeric substances will be influenced in some way by exposure to a plasma environment. The nature of the resulting changes in the polymer surface is essentially dependent upon the plasma conditions, including the nature of the gases introduced into the glow discharge. In most published cases, enhancement of wettability of and/or adhesion to such treated surfaces was observed. If one considers the nature of glow discharge (chapters 2 and 3) as well as the data shown in this chapter, it appears that one is justified in stating the following postulate:

> There exists a plasma environment for most, if not all, polymers that results ultimately in the eventual enhancement of the quality of adhesive bonds.

Indeed, a wide heterogeneity of polymer surfaces has been investigated and has been shown to exhibit marked enhancement in adhesion and adhesion-related properties.

Thus, Boenig and Blenner [78] demonstrated more recently that helium-plasma treatment of glass-reinforced cured-epoxy specimens results in up to 233% improvement of adhesive bonds to vulcanized rubbers (table 7.16). The plasma-treated samples failed because of cohesive rupture. Therefore neoprene, being an excellent adhesive, could not be markedly improved by plasma beyond cohesive failure.

An intriguing plasma treatment that included application to practical problems was disclosed in U.S. patent 4,123,308 [79]. Vapor-deposited para-xylylene polymers (parylenes) are commonly employed to coat or encapsulate various types of substrates, since they are insoluble in common solvents at room temperature, are scratch- and moisture-resistant, and exhibit low permeability to most gases and vapors. These polymers have also been found to remain tough and flexible over a wide temperature range and are therefore

Table 7.16. Effect of He-plasma on bond strength of rubber to epoxy-glass composite

Uncured rubber, cured to composite	Film adhesive #3351–1 PLI	Film adhesive #3351–1 and plasma-treated composite PLI	Improvement (%)
Natural rubber A 135 P	531	1116	110
Nitrile rubber K 135 A	305	1015	233
Neoprene rubber N 135 A	1055	1178	12

Source: H. V. Boenig and D. R. Blenner, unpubl. data, Lord Kinematics (1979).

suitable for use in coating of electronic assemblies such as printed circuit boards.

Sometimes it becomes necessary to repair parylene-coated electronic assemblies. This involves removing the parylene coating around the defective component and replacing the component. Then a patching compound, typically a thermosetting resin, is applied around the new replacement component to protect it. This patching compound must adhere to the repaired region and the unremoved parylene coating around the perimeter of this region. An important problem encountered in the commercial application of thermosetting resins has been the difficulty in making them adhere to the parylene surface to be coated. The aforementioned invention gives evidence that treatment of the parylene surface by oxygen plasma in a confined region imparts adhesive bonds that proved to be equal to the cohesive strength of the parylene itself. A typical example of the effectiveness of such plasma exposure to cured polyurethane thermoset resin is illustrated in table 7.17.

Similar results were obtained with poly(monochloro)-p-xylylene. It can be seen from table 7.17 that adhesive bonds of over tenfold strength were obtained by such plasma treatment. Furthermore, it is interesting to note that when the polyurethane thermosetting resin was applied to the plasma-treated parylene surface, the adhesive strength was found to be not only undiminished but actually enhanced (table 7.18).

Although there is a large volume of published data on grafting chemical entities to solid surfaces via high energy radiation, very little attention has been given to the use of plasma in adhesion. Most published work deals with the grafting of polymers onto the surface of fibers, metals, or other inorganic materials.

Three methods of plasma grafting have been described. In one, the substrate is exposed to a discharge sustained in the gas or vapor of the monomer to be grafted. The weakness of this approach is that the monomer forms a

Table 7.17. Effect of plasma exposure of parylene on the adhesive strength to a cured polyurethane resin

Plasma	Exposure time (min)	Adhesion (psi)
None	0	26
Oxygen	1	180
Oxygen	6	200
Argon	1	220
Argon	6	390
Helium	1	190
Helium	6	355

Source: T. E. Nowlin and R. A. Martineau, U.S. Pat. 4,123,308 (1978).

Table 7.18. Effect of layover time on adhesion

Plasma gas	Adhesion (psi) immediately after plasma treatment	One week after plasma treatment
Air	233	322
Oxygen	178	268
Argon	319	365
Helium	353	408

Source: T. E. Nowlin and R. A. Martineau, U.S. Pat. 4,123,308 (1978).

polymer in the plasma without attaching itself to the desired surface or forming only a weak bond with the substrate.

Another potentially more successful approach is to expose only the substrate to the glow discharge. In this way free radicals would be generated on the surface that can be employed for grafting in three ways. In one the treated surface is contacted immediately after removal from the plasma with the monomer, either in the liquid or vapor phase. The surface free radicals are then available for chain initiation. This approach requires precautions to prevent a massive decay of the surface free radicals. By this method polymerization can start only at the surface of the plasma-treated material, and it should provide a better graft. In a second approach, some homopolymerization may take place by this method because of a chain transfer mechanism, but the extent should be much less than by the first method.

A third approach is to expose the substrate to be treated to air or oxygen plasma in such a manner as to generate peroxides on the surface. The use of peroxides as catalysts in free-radical polymerization is known. Such treated surfaces could be exposed to suitable monomers at a later time under the most effective conditions.

The grafting in a corona discharge of a number of acrylates to cellulose was reported quite some time ago [80]. More recently, acrylonitrile was grafted to cotton fiber in an r.f. discharge of argon [42] with the formation of only negligible amounts of homopolymer. Difluoroethylene and tetrafluoroethylene were also reported to be successfully grafted onto cotton to obtain a hydrophobic surface.

Auerbach and O'Connor [40] disclosed a treatment in argon plasma of a Kevlar (polyaramide) fabric, followed quickly by immersion of the treated fabric in epoxy resin. In this way the investigators obtained an increase of the cured fabric in interlaminar shear strength of approximately 45% over the untreated material. It appears that in this technique the free radicals formed on the fibers reacted with the epoxy resin to form covalent bonds with the resin.

A similar approach was disclosed in a U.S. patent [81] in which the bond-

Table 7.19. Effect of plasma gas on peel adhesion of tire cord to rubber

Cure condition		Strip adhesion peeling load at 120°C (psi)						
Temperature	Exposure time	N_2	Ar	He	NH_3	Air	O_2	Control
210°C	3 min	16.9	14.5	11.5	21.0	—	7.5	2.5
232°C	3 min	18.6	17.5	10.6	23.7	14.1	11.2	3.0

Source: E. L. Lawton, U.S. Pat. 3,853,657 (1974).

ing of polyester tire cord to rubber was improved by brief exposure to a nonoxidizing low-temperature gas plasma. Extensive efforts have been made in the past toward the improvement of the bonding characteristics of such textile materials as rayon, nylon, and polyester to rubber for the manufacture of tires. In the case of certain synthetic fibers, due to their surface characteristics, the resin-latex composition commonly employed as the adhesive has generally not been satisfactory in holding the fiber side of the bond.

In the case of polyester, other surface treatment agents such as isocyanates have been used in conjunction with the resin latex composition, as a pretreatment for the fiber. Handling and safety precautions in the use of isocyanate made it desirable to search for other fiber treatment methods, however.

The method described in this patent used bundles of 1,000 denier/200 filament polyethylene terephthalate fibers, which were processed through a commercial plasma generator at 0.5 mm Hg for 9.4 seconds using nitrogen and oxygen gases at a flow rate of 20–40 ccm per minute. A standard RFL dip (resorcinol formaldehyde latex) was subsequently applied, and the treated cord passed through a 6 ft drying zone at 165°C before bonding to rubber. The effect of plasma treatment in various gases on the peel adhesion of the test strips (cord adhesives/rubber) is shown in table 7.19. Evidently, adhesive bond strength was significantly improved until it was nearly eight times that of the untreated control.

Finally, mention should be made of the industrial success of the plasma treatment of nonpolar surfaces, such as polyolefins, both to enable printing on these surfaces and to obtain adequate adhesion of the prints. In these cases the object, such as a writing pen, is exposed to an oxygen-containing plasma for 2–3 min and is then imprinted by some process [82].

3. Abrasion Resistance and Wear

It has generally been observed that most plasma polymers are highly crosslinked and are therefore harder than their conventional analogs. There are also several patents in existence that claim methods for improving the abrasion resistance of polymer surfaces by either plasma exposure or plasma deposition.

A recent U.S. patent [83] describes a plasma-coating composition based on a group of organo-silane compounds that combines high abrasion resistance, it is claimed, with optical clarity. Specific mention is made of the following compounds: vinyltrichlorosilane, tetraethoxysilane, vinyl-triethoxysilane, tetravinylsilane, vinyltriacetoxysilane, hexamethyldisilazene, tetramethylsilane, vinyldimethylethoxysilane, vinyltrimethoxysilane, and methyltrimethoxysilane.

It is shown in this disclosure that abrasion resistance for plastic optical surfaces was improved from 100 plays to over 1200 plays without visual degradation and also provided an optically clear layer in the range of 4,000 Å to 8,000 Å. It is noted that the claims include plasma coatings from silane monomers that impart protection against wear to polymers such as acrylics, polycarbonates, and polystyrenes. The coating compositions were found to be highly compatible with the plastic substrate.

This process was recently modified [84] for application to plastic lenses to include a first-step exposure to a water-vapor plasma to enhance adherence of the deposited film. A second step deposits the silane-based coating, and a third step entails argon-plasma exposure to increase crosslinking and thus scratch resistance.

The abrasion resistance of a polycarbonate coated with plasma-deposited vinyltrimethoxysilane was tested in one reported case using a Tabor Abrasor to which a hazemeter was attached. The ratio of the scattered light to the total light was taken as a quantitative measure of abrasion [85]. The results of these experiments are listed in table 7.20. The smallest amount of damage from abrasion is shown to occur in coated specimens that had been posttreated in an oxygen glow discharge. They show a haze value only 0.2% greater than that of uncoated, unabraded polycarbonate.

An English patent [86] relates increased wear resistance to hardening of a polymer surface by plasma-induced crosslinking of a dielectric coating of

Table 7.20. Hazemeter measurements of abrasion resistance of plasma coatings from vinyltrimethoxysilane

Polycarbonate specimens	Haze (%)	Number of tests averaged
Uncoated	5.2	10
Coated	4.1	3
Coated and oxygen treated	2.5	6

Source: J. C. Fletcher, T. J. Wydeven, Jr., and J. R. Hollahan, U.S. Pat. 4,137,365 (1979).

poly-*p*-xylylene for recording and playback systems. According to one method, conductive video discs are prepared having geometric variations in the bottom of a spiral groove in the disc surface that correspond to capacity variations representative of the stored information. The conductive disc is prepared by metallizing the surface of a vinyl disc. These discs are then coated with a thin conformal dielectric coating. A stylus with a metallic electrode completes the capacitor and, during playback, rides upon the dielectric coating, detecting the variations in the groove. It is therefore desirable that this coating not only be smooth and pinhole-free but also exhibit good wear resistance to withstand frequent passes of the stylus. A conformal dielectric coating of poly-*p*-xylylene was found to meet all but one of the stringent requirements; the coating is too soft, being capable of providing only about 100 plays of satisfactory quality. The same patent [86] relates to the hardening of the dielectric poly-*p*-xylylene coating by exposing the coated disc to a 20 kHz air glow discharge for over 30 sec. The wear quality of the disc was thus reported to be improved from 100 plays to over 1200 plays without visual degradation.

4. Hardness

Although ultrathin films as frequently obtained in plasma deposition are not suitable for accurate hardness measurements, resistance to indentations has been occasionally used as a qualitative comparison test. In most reported cases increased hardness has been related to high crosslinking and/or oxidation and has also been compared with improvements in abrasion resistance.

Aisenberg and Chabot [87] deposited diamondlike films employing an ion beam of carbon produced in an argon plasma followed by introduction into a low-vacuum vessel. The transparent films were reported to show evidence of extreme hardness and properties that resemble those of diamond. Other investigators [88,89] showed that hard films can be obtained by cracking a hydrocarbon gas in a d.c. or r.f. glow discharge. The substrate is exposed alternately to positive ions and electrons on a negatively biased electrode in a plasma and thus prevents any net accumulation on insulating deposits. Under optimum conditions no IR absorption bands due to C—H have been found, and at higher power, evidence of graphitization was detected.

Swedish researchers [90] commented more recently on the structure of carbon film produced by cracking butane in a r.f. glow discharge for 2 to 5 min at 150–300 watts. All films were reportedly very hard and had good adhesion to the glass substrate. The films showed the spectrum for diamond and those exposed to higher power exhibited the spectrum of microcrystalline graphite. The authors suggest that films produced at low deposition rates have similarities to bulk diamond, although they can in no way be called diamond.

5. Friction Characteristics

In a recent study [91] fluorocarbon films were prepared on slides by (a) r.f. sputtering of polytetrafluoroethylene in argon glow discharges and (b) plasma polymerization of CF_4 gas on the substrates. The coefficient of static friction was measured for glass and steel riders on the fluorocarbon deposits using the method of detecting movement of a rider on a plane of variable inclination. The friction coefficient for the films was between two and four times that of the bulk polymer irrespective of the mode of film formation, the rate of growth, and the thickness in the range 300–26,000 Å. Generally, plasma-polymerized films were harder than sputtered coatings. Some of the results of this study are listed in table 7.21.

On the other hand, enhanced lubricity was claimed by Auerbach in a U.S. patent [92] by plasma-treating inorganic and organic substrates in a difluorocarbene or trifluoromethyl reactive species. It has earlier been shown [93] that the critical surface-tension values of several polymers decrease with increasing exposure time of difluorocarbenes in a glow discharge. It is assumed that the reaction involves insertion of $:CF_2$ into the surface of a hydrocarbon polymer to form a surface group

$$\boxed{}\!\!\!-\!\!\!-H + :CF_2 \rightarrow \boxed{}\!\!\!-\!\!\!-CF_2H$$

thus imparting surface-tension values approaching those of polytetrafluoroethylene or of other fluorinated polyethylenes (table 7.22).

Table 7.21. The relative static friction coefficient μ_R as a function of the power and thickness of the films

	Power (W)	Thickness (Å)	Steel rider μ_R	Glass rider μ_G
PTFE sputtered in Ar	100	5000	3.63	3.39
	200	7400	3.34	2.88
	200	26400	3.42	3.59
	300	7000	2.8	2.56
PTFE sputtered in its own vapors	100	340	3.51	3.59
	200	2100	3.63	4.62
	300	3000	3.26	3.59
Plasma-polymerized films	100	550	4.44	—
	200	2000	4.62	—
	200	6000	3.75	—
	200	8800	3.27	—
	300	3000	2.35	—

Note: $R = \mu_F/\mu_B$ where μ_F and μ_B are the static friction coefficients for film and bulk PTFE, respectively, and μ_B is 0.13 and 0.195 for steel and glass riders, respectively.
Source: H. Biederman, S. M. Ojha, and L. Holland, *Thin Solid Films, 41,* 329 (1977).

Table 7.22. Critical surface tensions of difluorocarbene modified polymer surfaces

Polymer	Time of exposure (min)	(Dynes/cm)
Polyethylene	0	31.0
	3	27.0
	8	23.5
	14	21.0
Polypropylene	0	32.0
	3	30.0
	8	27.0
Polymethylmethacrylate	0	39.0
	3	35.0
	8	31.0
Nylon	0	41.0
	8	34.5
	14	24.5
Polystyrene	0	30.5
	8	24.0
Cellulose acetate	0	39.0
	3	26.0
	8	20.0

Source: D. A. Olsen and A. J. Osterhaas, *J. Appl. Polym. Sci., 13,* 1523 (1969). Copyright © 1969, John Wiley & Sons, Inc. Reprinted by permission of John Wiley & Sons, Inc.

6. Diffusion Characteristics

In section 2B of this chapter the effect of CASING on adhesion has been discussed. Crosslinking was shown to occur if polymers are exposed to a glow discharge of an inert gas, such as argon. It has also been shown in chapter 5 that highly crosslinked polymer films are obtained by plasma deposition on a variety of surfaces.

Bell and coworkers [94] have reported on interesting diffusion experiments on both types of surfaces. One application was aimed at the reduction of diffusion or leaching of low-molecular-weight compounds from polymers employed in prosthetic implants. There has been evidence that the exudation of materials such as plasticizers, antioxidants, and initiators can be harmful to the body of the host. These investigators [94] determined the leaching of low-molecular-weight material by sealing the sample in an ampule containing a simulated body fluid (PECF) followed by autoclaving at 115°C and a pressure of 31 psi for 62 hr. The amount of extracted material was isolated and determined by IR spectroscopy. Figure 7.13 illustrates the effect of CASING in an argon plasma on the fraction ϕ that was leached out by PECF. ϕ is defined as the amount of low-molecular-weight material leached from the

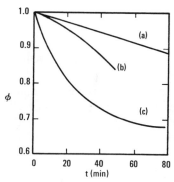

Figure 7.13. Fractions (ϕ) of impurities leached from polymers treated by CASING in argon as a function of plasma exposure time. (*a*) Polypropylene; (*b*) Poly(ethylene terephthalate); (*c*) Poly(methyl methacrylate). *Source:* Adapted from F. Y. Chang, M. Shen, and A. T. Bell, *J. Appl. Polym. Sci., 17,* 2915 (1973). Copyright © 1973, John Wiley & Sons, Inc. Reprinted by permission of John Wiley & Sons, Inc.

plasma-treated polymer divided by that leached from the untreated polymer. It is evident from figure 7.13 that CASING produced an increasingly impenetrable surface as the exposure time to argon plasma was increased, with poly-(methyl methacrylate) being the most effective material in this respect.

The effect of plasma deposition in an ethylene glow discharge on leaching is shown in figure 7.14. Deposition of a highly crosslinked plasma polyethylene film is shown to be more effective in sealing off polymer surfaces than is a CASING process. The investigators [94] point out that the effectiveness of the treatment depends on the discharge operating conditions. At 100 watts and 2.0 torr, a 60-min treatment resulted in a 95% reduction in the extent of extraction, while at 80 watts and 1.8 torr, the same treatment is only 50% effective.

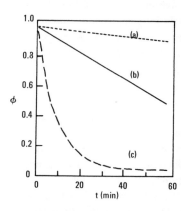

Figure 7.14. Fractions (ϕ) of impurities leached from polymers treated by plasma deposition of ethylene. (*a*) Poly(vinyl chloride), 100 watts, 2 torr; (*b*) Poly(ethylene terephthalate), Poly(methyl acrylate), both 80 watts, 1.8 torr; (*c*) Poly-(methyl acrylate), plasticized with dioctyl phthalate, Poly(ethylene terephthalate), all 100 watts, 2 torr. *Source:* Adapted from F. Y. Chang, M. Shen, and A. T. Bell, *J. Appl. Polym. Sci., 17,* 2915 (1973). Copyright © 1973, John Wiley & Sons, Inc. Reprinted by permission of John Wiley & Sons, Inc.

It is interesting to note that the sealing effect by plasma deposition, while very remarkable on poly(methyl acrylate) and poly(ethylene terephthalate) at higher power, is practically nonexistent on poly(vinyl chloride).

7. Permselectivity and Reverse Osmosis

A. Introduction

The permselectivities of ionic and nonionic polymer membranes have been more intensely investigated in recent years. For industrial applications, membranes of high selectivity and permeability have found broad appeal among researchers. This activity has led to novel and potentially promising methods for preparing reverse osmosis and permselective membranes that involve the use of plasma produced in an electrical discharge to deposit a thin polymer layer on the surface of a porous substrate [95,96,97]. Solute rejection or molecular selectivity is accomplished by the plasma-deposited film.

There are many advantages of this technique over conventional casting methods, such as preparation and storage of the finished membrane in a dry state, independent selection of substrate and film composition, and the ability to prepare an ultrathin rejection layer, thereby forming membrane systems that permit high water fluxes.

Distinction is generally made between membrane selectivity in separation of gas mixtures, also sometimes referred to as gas permselectivity, and that in separation of ions and water, such as for desalination purposes, generally referred to as reverse osmosis mechanism.

Beyond these two general application areas it is generally recognized that the combination of extremely thin films and the layered combination of such films having different selectivities in terms of molecular size, solubility, ion affinity, and diffusivity renders research in this field most intriguing for a variety of separation problems, such as desalination of seawater, purification of drinking water, concentration of alcohol solutions in gasohol applications, purification in artificial kidneys, and air purification, especially in confined spaces, such as submarines, space capsules, and oxygen tents.

B. Permselectivity for Gases

When using unmodified, commercially available polymer films for separations of interest, it is frequently observed that if selectivity is at the desired level, the permeability is too low, and vice versa. This behavior is understood from the polymer structural features associated with selectivity and permeability. Selectivity is enhanced as the differences in solubility in the membrane are increased and as the spacing between polymer chains is small enough and restricted enough to distinguish between diffusing molecules of different sizes. These polymer structural features for high selectivity would require inflexible polymer molecules that are further constricted in their efficiency by the presence of crystallites, crosslinking, hydrogen bonding, or high van der Waals forces. These same features, however, are impediments to permeability.

In the past, one approach to solving this problem was to use a film of high selectivity but extreme thinness, that is, about 0.5 microns, to attain high diffusion rates at reasonable pressures or concentrations. Until recently, however, pinhole-free films below 5 microns could not be produced. The most successful approach, prior to the plasma-membrane era, was the technique, introduced in 1962, of overlaying a higher permeable thin film onto an ultrathin selective structure [98].

Ten years later, Stancell and Spencer [99] prepared plasma-deposited films on the order of 0.5 micron thickness on a silicone-carbonate copolymer substrate and then measured effects on selectivity and permeability for gas mixtures, such as hydrogen and methane. The permeability, P, was derived from the relationship:

$$P = \frac{J \cdot l}{\Delta p} \qquad \text{(Eq. 7.12)}$$

where l is the film thickness and Δp is the total pressure drop across the film. Thus, the hydrogen-to-methane permeability ratio for the untreated silicon-carbonate film was 0.87, that is, the methane permeability was higher than the hydrogen permeability. After plasma deposition, using benzonitrile monomer, the ratio was increased to 33.0 at just a 20% loss in hydrogen permeability (table 7.23).

The ratio of less than unity for the untreated film apparently indicates that the larger size of the methane molecule, implying lower permeability, is offset by its higher level of dispersive power, favoring higher solubility. The average collision diameter is 3.8 Å for methane and 2.9 Å for hydrogen; however, methane is 5 to 10 times more soluble in most organic solvents than is hydrogen.

The considerably higher selectivity of the plasma-deposited films to hydrogen permeation suggests that such thin top coatings are capable of distinguishing between hydrogen and methane on the basis of molecular size. The primary elements of a system for gas permeability are illustrated in figure 7.15.

It has also been reported [100] that coatings of polypropylene formed in a r.f. plasma on polyethylene film cause a reduction of the permeability of the gases, such as He, Ne, O_2, and CO_2. In the cases of O_2 and CO_2, the solubility of the gases was found to be decreased, the diffusion behavior being the predominant mechanism. This behavior could be important in those applications where a definite ratio of two gases is desired.

C. Reverse Osmosis

Yasuda and his coworkers [101,102] have demonstrated that salt rejection and water throughput of a homogeneous polymer membrane are related and

Table 7.23. Hydrogen and methane permeabilities of plasma films deposited on silicon-carbonate copolymer films

Monomer	Permeability $\times 10^7$ at 25°C cc (S.T.P.)-cm/cm²-sec-atm		Permeability ratio, H_2/CH_4
	H_2	CH_4	
None	9.2	10.5	.87
Nitrile type			
Cyanogen bromide	2.74	.075	36.6
Nicotinonitrile	3.96	.11	36.0
Benzonitrile	7.26	.22	33.0
3-Butenenitrile	8.87	.39	21.7
Methacrylonitrile	9.04	.55	16.4
Chloroacetonitrile	13.0	4.97	2.6
Acrylonitrile	15.3	4.37	3.5
Vinyl type			
Acrylic acid	8.15	.33	24.5
Allyl bromide	8.37	.66	12.6
Styrene	8.46	.68	12.4
Divinylbenzene	9.85	1.33	7.0
Aromatic			
Naphthalene	8.1	.44	18.4
Benzene	9.57	1.33	7.2
Mesitylene	9.43	1.72	5.5
1,3-Di(trifluoromethyl)-benzene	9.92	5.98	1.7

Source: A. F. Stancell and A. T. Spencer, *J. Appl. Polym. Sci., 16,* 1505 (1972). Copyright © 1972, John Wiley & Sons, Inc. Reprinted by permission of John Wiley & Sons, Inc.

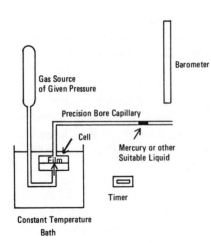

Figure 7.15. Elements of system for gas permeability measurements. *Source:* Adapted from A. F. Stancell and A. T. Spencer, *J. Appl. Polym. Sci., 16,* 1505 (1972). Copyright © 1972, John Wiley & Sons, Inc. Reprinted by permission of John Wiley & Sons, Inc.

that satisfactory salt rejection can be obtained only with a waterflux constant below a certain threshold value. They subsequently observed [96] that for high-throughput membranes, the polymer in the rejecting layer should be moderately hydrophilic. As a class, only nitrogen-containing compounds, such as olefinic, heteroaromatic, and aromatic amines were found to form polymer films that exhibited satisfactory reverse-osmosis characteristics.

The minimum thickness of polymer deposition for a flawless membrane is the radius of the pore it must bridge. Therefore, the efficiency of reverse osmosis of a composite membrane depends largely on the pore size of the porous substrate (table 7.24). The authors point out [96] that the best method for polymer deposition of such membranes depends on the type of substrate; for instance, the best method for Millipore filter is not necessarily best for porous polysulfone. It is shown in table 7.25 that good results are obtained from plasma-deposited 4-vinylpyridine.

The reverse osmosis membranes of plasma polymers are said [96] to have unique features that are contrary to commonly observed trends:

Steady increase of salt rejection and water flux with time of reverse osmosis run.
Very stable water flux after an initial incline.
Excellent performance at high salt concentration and under high pressure.

Yasuda and colleagues [101] observed that addition of N_2 and H_2O in the plasma deposition of membranes from 4-picoline, 3,5-lutidine, benzene, and acetylene improves the reverse osmosis characteristics, especially water flux, of the polymers to a significant extent. The addition of nitrogen gas into the plasma of monomers that do not contain nitrogen atoms was found to produce plasma polymers similar to those formed from nitrogen-containing monomers. The addition of water vapor increases the stability of the mem-

Table 7.24. Reverse osmosis characteristics of porous substrates

Porous substrate	Thickness (μ)	Pore size (Å)	Water flux (gfd)	Salt rejection (%)
Millipore filter VS	48	250	400–500	0
Polysulfone	13	—	300–400	2–4
Porous glass	1100	50	0.6–1.0	5–10

Notes: Water flux and salt rejection were measured with 1.2% NaCl solution at 1200 psi applied pressure. Figures cited were taken after the initial drop of water flux, which generally occurred in 2–3 hr, had been observed and the flux had stabilized.
Source: H. Yasuda and C. E. Lamaze, *J. Appl. Polym. Sci., 17,* 201 (1973). Copyright © 1973, John Wiley & Sons, Inc. Reprinted by permission of John Wiley & Sons, Inc.

Table 7.25. Reverse osmosis results of plasma-deposited 4-Vinylpyridine membranes with various porous substrates

Substrate	Salt rejection (%)	Water flux (gfd)[a]
Porous glass		
17	90	0.49
23c	96	0.81
Polysulfone		
Ro 165	79	9.9
Ro 169	89	4.0
EPS 1	86	26.4
Millipore VSW		
Ro 12	83	17.0
Ro 14	99	38.0
Ro 91	89	2.0
Ro 98	96	1.1
Ro 105	97	7.4

Note: 1.2% NaCl at 1200 psi.
[a]gfd = gal/ft² day.
Source: H. Yasuda and C. E. Lamaze, *J. Appl. Polym. Sci., 17,* 201 (1973). Copyright © 1973, John Wiley & Sons, Inc. Reprinted by permission of John Wiley & Sons, Inc.

branes formed; this seems to coincide with a considerable reduction of free radicals found in those polymers.

It has been shown by Bell and colleagues [103] that in order to obtain membranes exhibiting high rejection, the microporous structure of the substrate must be completely covered by the plasma-deposited polymer. The latter should be as thin as possible so as to ensure high water flux. These

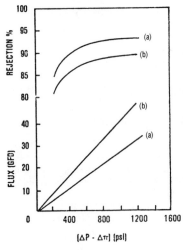

Figure 7.16. Effects of applied pressure on flux and rejection. Membrane substrate: (*a*) 432, Gulf 3; (*b*) 409, Gulf 2. *Source:* Adapted from A. T. Bell, T. J. Wydeven, and C. C. Johnson, *J. Appl. Polym. Sci., 19(1),* 911 (1975). Copyright © 1975, John Wiley & Sons, Inc. Reprinted by permission of John Wiley & Sons, Inc.

investigators also studied the performance of several membranes as a function of the applied pressure (figure 7.16). In this figure $\Delta p - \Delta \pi$ represents the difference between the applied pressure and the net osmotic pressure between feed and effluent.

Yasuda and Lamaze [101] have reported that the salt rejection of a water-swollen membrane can be expressed by

$$R_s = \{ w + [P_2 RT/P_1 V_1 (\Delta p - \Delta \pi)] \}^{-1} \qquad \text{(Eq. 7.13)}$$

where R = salt rejection, P_1 = diffusive permeability of water, P_2 = diffusive permeability of salt, and V_1 = the molar volume of water. The parameter ω is defined by

$$\omega = K_1 RT/P_1 V_1 \qquad \text{(Eq. 7.14)}$$

where K is the hydrolic permeability. For this model, the total water flux is given by

$$J_1 = K_1 (\Delta p - \Delta \pi) /\Delta x \qquad \text{(Eq. 7.15)}$$

where Δx = thickness of either type membrane (for a homogeneous membrane) or the rejecting layer (for a composite membrane).

From equation 7.13 it follows that as $(\Delta p - \Delta \pi)$ approaches infinity, R_s approaches ω^{-1}. If water flow through the membrane occurs by diffusion only, $\omega = 1$ and R_s has a limiting value of 1. When water flow takes place by both diffusive and bulk transport, $\omega > 1$ and $R_s < 1$.

Bell and colleagues [103] demonstrated recently that for pressures above 600 psi, their membranes formed from plasma-deposited allylamine obey equation 7.13 and that most of the water flow through the membrane occurs by diffusion and is to a lesser extent due to bulk flow.

8. Ion-Exchange Properties of Plasma Membranes

Ion-exchange resins are known to be crosslinked polyelectrolytes. Crosslinked polystyrene is particularly suitable for the synthesis of ion-exchange resins because the introduction of various ionically dissociating groups into the aromatic ring can be readily accomplished. Thus, reaction with SO_3 produces a strongly acidic cation-exchange resin, as shown below.

$$-CH_2-CH-$$

$$SO_3H$$

The parent structures are modified so that these resins swell in water to afford the dissociating groups greater accessibility. The exchange of the dissociating, low-molecular-weight ions of the resins is an equilibrium reaction, so that water-containing salt can be completely freed of salt by passing through a cation- and anion-exchange resin with the polyions, (poly)$^-$ or (poly)$^+$, respectively.

$$(poly)^- \; H^+ + Na^+ \rightleftharpoons (poly)^- \; Na^+ + H^+$$
$$(poly)^+ \; OH^- + Cl^- \rightleftharpoons (poly)^+ \; Cl^- + OH^-$$

$$(poly)^- H^+ + (poly)^+ OH + NaCl \rightleftharpoons (poly)^- Na^+ + (poly)^+ Cl + H_2O \quad (Eq. \; 7.16)$$

The spent ion-exchange resins can then be regenerated by acid or alkali treatment. The use of such cation resins as membranes for electrolytic cells is becoming increasingly important to the chemical industry. This development is chiefly due to their greater ecological acceptability, potential for energy conservation, and the cost reduction that they are claimed to bring about.

One major element for the hardware aspect of the electrolytic system is the cation-exchange membrane that separates the anode compartment from the cathode compartment within the electrolytic cell to provide a divided electrolytic cell for more efficient electrochemical production. At this time, the membrane having greatest utility is one that can be employed in a chlorine and sodium hydroxide cell, because in this country chlorine and NaOH are produced almost exclusively by electrolysis from aqueous solutions of sodium chloride. Both of these chemicals, being large-volume commodities, are to a large extent manufactured from diaphragm-type electrolytic cells. In this process, sodium chloride solutions are continuously fed into the anode compartment and flow through the diaphragm backed by a cathode, while hydrogen ions are discharged in the form of hydrogen gas. If the membrane is capable of producing at high current efficiencies and can resist the corrosive anolyte solution consisting of free halide, it is believed that a significant purity and concentration increase in the end product should be possible and could save secondary steps in such a process.

In a recent patent McCain [104] claims an improved hydrolically impermeable cation-exchange membrane composite, which is coated by plasma deposition using monomers, such as tetrafluoroethylene or a mixture of acetylene, nitrogen, and water. The results are compared with the untreated membrane composite in table 7.26. It is shown that the current efficiency at comparable voltages and current densities of the plasma-treated membrane composite is approximately 35% greater than that of the untreated material and that it exhibits good lifetimes. Therefore, it would seem that this general approach would be met with considerable future research interest.

Table 7.26. Effect of plasma coating using TFE of cation exchange membrane composite

	Untreated	Plasma treated
Gel Water after 10-hr boil (%)	20–25	19.2
Current efficiency (%)	53–62	70–81.5
Duration in cell without failure (days)	200	56

Source: Compiled from H. McCain, U.S. Pat. 4,100,113 (1978).

9. Electrical Properties

A. Introduction

Due to their excellent dielectric characteristics, high-polymer materials are widely employed in the manufacture of miniature capacitors. Both polystyrene and polyethylene terephthalate stretched films have been used frequently as thin-film dielectrics. In the field of microelectronics, however, no completely satisfactory manufacturing techniques for applying ultrathin films of satisfactory uniformity and integrity were found until the application of plasma-deposition technology in the late sixties. The advantages of such thin-film fabrication techniques combined with the noncrystalling nature of plasma-deposited polymers and the desired crosslinked structure have generally been described as factors in the relatively rapid acceptance of this technology in the seventies.

B. Dielectric Constant

The values of the dielectric constants at 10^6 cycles of commercial polymers may range from 2.0 for polytetrafluoroethylene to 6.2 for cellulose acetate butyrate. When filled with appropriate powders, these values may go as high as 10–15.0 for mineral containing cast phenol formaldehyde resins.

Pinhole-free organic insulators are generally used for electrical insulation applications, such as capacitors. One major problem observed in such insulators is electrical degradation over a period of time, which is frequently caused by corona at the interface between the insulator and the environment. The corona, promoted by the large electric field at the surface, increases the decomposition of the organic dielectric. The large surface electric field is induced by the large discontinuous change in the dielectric constant across the insulator-gas or insulator-liquid interface. Organic insulators have large dielectric constants (mentioned above), while gases have values near 1.0 and values of liquids range from 2.5 to 6.5. The area of a discontinuous change in the dielectric constant is frequently observed to be the point of corona occurrence.

In order to prevent decomposition and ultimate electric failure of the insulator, it is necessary to effect a reduction in the operating electric field at the surface. If the insulator were coated with a dielectric film that would eliminate the discontinuous change in the dielectric constant, then decomposition of the insulator could be significantly retarded (figure 7.17 a and b).

Figure 7.17. Effect of dielectric coating on discontinuous changes in (*a*) dielectric constant and (*b*) relative electric field. *Source:* Adapted from M. Hudis and T. Wydeven, U.S. Pat. 4, 132, 829 (1979).

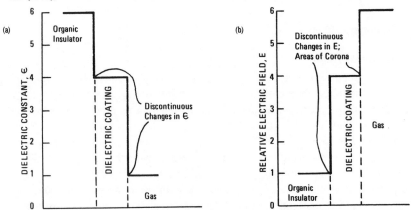

A recent U.S. patent discloses [105] that the dielectric constant of a polymer coating can be varied and controlled by a plasma-polymerization process in which the current density and electron energy, both of which are related to gas pressure and to electric field strength and frequency, are varied and controlled. It is claimed that this eliminates the discontinuous change in the dielectric constant across the interface and thereby eliminates the formation of corona. The monomers that can be used in this process to prepare coatings having high dielectric constants contain polar groups in their structure, such as nitrogen-containing amines and nitriles. Monomers not endowed with such polar groups can also be used in such a plasma deposition by adding

Table 7.27. Effect of plasma parameters on dielectric constant of polymers from allylamine

Monomer pressure (torr)	Temperature (°C)	Current density (A/cm² Rms)	Dielectric constant
0.06	20	20	5.5
0.06	20	113	6.8
1.35	68	300	7.1

Note: A.c. electric field frequency = 10 kHz.
Source: M. Hudis and T. Wydeven, U.S. patent 4,132,829 (1979).

Table 7.28. Effect of plasma parameters on the dielectric constant of films obtained from a mixture of ethylene and nitrogen

Partial pressure ethylene (torr)	Partial pressure nitrogen (torr)	Temperature (°C)	Current density (μA/cm² Rms)	Dielectric constant
0.2	0	20	1700	3.3
0.2	0.2	20	5650	4.1
0.2	0.2	20	6280	5.9

Source: M. Hudis and T. Wydeven, U.S. patent 4,132,829 (1979).

nitrogen to the monomer in the reaction chamber, which is then incorporated in the plasma-polymer structure in the form of polar groups. The effect of varying plasma parameters on the dielectric constant of deposited films from allylamine is shown in table 7.27.

When a mixture of ethylene and nitrogen was plasma-deposited between electrodes at 13.56 MHz, the dielectric constant of the resultant nitrogen-containing film was parameter-dependent, as shown in table 7.28.

It appears from these data that an increase in current density facilitates the introduction into the forming polymer matrix of additional polar nitrogen-containing groups.

C. Capacitance

It has early been reported [106] that uniform films obtained by plasma polymerization of tetrafluoroethylene have very few defects and can provide up to at least 300 pF/mm² capacity (figure 7.18). The thickness of the film was calculated from a dielectric constant of 2.0 and 170 pF/mm² to be 0.1μ. Heat exposure of the film for 2 hr at 150°C caused no change in capacitance; neither did the capacitance of such deposits change with the frequency. Furthermore, such films displayed no significant change in capacitance when 20 d.c. volts were applied and the films were aged at 80°C for 1000 hours. The importance of the disclosure of these data in 1969 [106] appears to lie in the fact that capacitors in the higher pF capacitance range became available for the first time through plasma-deposition technology.

In more recent times, Ristow [107] was the first to draw attention to changes in electrical properties of films when they absorb humidity. Such information is evidently of importance when such films are used as dielectrics or for coating electronic elements. He [107] plasma-deposited at r.f. frequencies films from fluorinated hydrocarbon monomers, such as fluorocyclobutane, hexafluoropropylene, perfluoromethyldecalin, and fluorobutylene, and determined the change of capacity after exposure to different levels of relative humidities. No swelling was observed under an interference micro-

Figure 7.18. Breakdown voltage versus capacitance of plasma PTFE. *Source:* Adapted from P. J. Ozawa, IEEE Trans. on Parts, Materials, and Packaging, PMP-5, No. 2, 112 (June 1969). © 1969 IEEE.

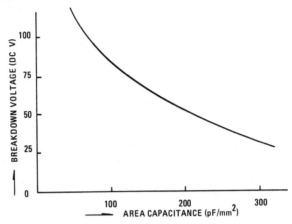

scope, and therefore the polar nature of the water was thought to be the cause for the measured increases in capacity.

The change of capacity of the plasma films in different relative humidity proved to be dependent on the dielectric volume only and not directly on the length of edge, the thickness, or the capacitive area. As a bulk property, moisture absorption has the same effect as changing the dielectric constant after absorption. If ϵ_1 and C_1 are the effective dielectric constant and the capacity for a dry sample of area F and thickness d, and ϵ_2 and C_2 the values after absorption, then the ratio

$$\frac{C_2 - C_1}{C_1} = \frac{\epsilon_2 \epsilon_0 \, F/d - \epsilon_1 \epsilon_0 \, F/d}{\epsilon_1 \epsilon_0 \, F/d} = \frac{\epsilon_2 - \epsilon_1}{\epsilon_1} \qquad \text{(Eq. 7.17)}$$

does not depend on the size of the sample. Therefore, if ΔC represents the change of the capacitance C due to a 1% change of relative humidity at constant temperature, then the ratio

$$\beta_c = \frac{\Delta C}{C} \text{ per 1\% relative humidity} \qquad \text{(Eq. 7.18)}$$

is a measure to describe the dielectric's property on moisture absorption. By definition, β_c is positive for an increase in capacitance with increasing relative humidity. The results for different monomers at two discharge frequencies are given in table 7.29. It is evident that the higher discharge frequency gives better results for all products. Surface tension, δ, as well as loss tangent values, are correlated with β_c.

Table 7.29. Electric properties of plasma-deposited films

Monomer	Discharge frequency (MHz)	$\beta_c \times 10^6$/rel. hum.	Tan δ (100 kHz) $\times 10^3$	6×10^5 (N/cm^{-1})
Perfluorocyclobutane	0.5	500[a]	1.9[a]	20[b]
	10	400[b]		
Hexafluoropropylene	0.5	600[a]	2.0[a]	20[b]
	10	400[b]		
Perfluoromethyldecalin	0.5	600[a]	2.8[a]	25[b]
	10	400[b]		
Perfluorobutylene	0.5	2200[a]	3.6[a]	30[b]
	10	600[b]	3.0[a]	

[a]Measured when the sample was in vacuum chamber immediately after fabrication.
[b]Measured after equilibrium had been reached under room conditions.
Source: D. Ristow, *J. Materials Sci., 12,* 1411 (1977). Reprinted by permission of the publisher, Chapman & Hall, Ltd.

D. Conductivity

Bradley and Hammes [108] measured conductivities of 30 plasma-deposited film specimens and the effect of temperature on these values. The results are illustrated in figure 7.19 for naphthalene and acrylonitrile. The accurate value for the film thickness, required for the calculation of conductivity, were obtained from capacitance measurements of the specimens as well as by counting interference fringes. The high thermal stability of most of these films permitted the measurement of the activation energy of conduction, E_a. The values of E_a were determined from the slopes of the log 6 versus $1/T$,

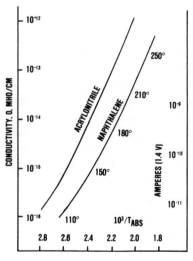

Figure 7.19. Experimental current and conductivity versus temperature for naphthalene and acrylonitrile plasma films. MHO = Conductance of a body through which one ampere of current flows when the potential difference is one volt. *Source:* Adapted from A. Bradley and J. P. Hammes, *J. Electrochem. Soc., 110*(1), 15 (1963). Reprinted by permission of the publisher, The Electrochemical Society, Inc.

Table 7.30. Conductivity values and activation energies for plasma polymers

Monomer	б mho/cm		E_a, eV
	150°C	250°C	
Naphthalene	9×10^{-16}	2.7×10^{-13}	1.1
Styrene	6×10^{-16}	9×10^{-14}	1.2
p-Xylene	5×10^{-17}	1.5×10^{-13}	1.8
Cyclopentadiene	1.0×10^{-16}	1.2×10^{-13}	1.5
Hexamethylbenzene	7×10^{-17}	7×10^{-14}	1.5
Ethylene oxide	4×10^{-16}	1.6×10^{-13}	1.1
Methoxynaphthalene	1.1×10^{-16}	7×10^{-14}	1.5
Thiourea	3.3×10^{-16}	4×10^{-13}	1.7
Chlorobenzene	8×10^{-17}	1.9×10^{-14}	1.4
Picoline	2.2×10^{-14}	6×10^{-12}	1.1
N-Nitrosodiphenylamine	8×10^{-15}	3.0×10^{-12}	1.2
p-Toluidine	7×10^{-16}	2.3×10^{-12}	1.5
Aniline	2.8×10^{-16}	1.4×10^{-12}	1.8.
p-Nitrotoluene	5×10^{-16}	2.5×10^{-13}	1.2
Diphenyl selenide	3.1×10^{-18}	8×10^{-13}	0.75, 1.5
Diphenyl mercury	2.8×10^{-15}	2.7×10^{-13}	0.85
Ferrocene	2.7×10^{-13}	4.5×10^{-12}	0.55
Benzene selenol	2.5×10^{-14}	7×10^{-12}	1.1
Hexa-n-butyl (di) tin	1.5×10^{-15}	7×10^{-13}	1.1
Tetracyanoethylene	1.8×10^{-13}	5×10^{-12}	0.60
Malononitrile	3×10^{-14}	1.8×10^{-12}	0.75
Thianthrene	1.5×10^{-14}	1.6×10^{-12}	0.85
Thiophene	6×10^{-14}	3×10^{-12}	0.75
Thioacetamide	8×10^{-14}	9×10^{-12}	0.85

Source: A. Bradley and J. P. Hammes, *J. Electrochem. Soc., 110* (1), 15 (1963). Reprinted by permission of the publisher, The Electrochemical Society, Inc.

where б is the surface energy. The conductivity data at 150° and 250°C as well as E_a values of a number of polymers measured are listed in table 7.30.

It is noted that all materials tested by these workers [108] that gave highly conductive films, such as plasma polymers from ferrocene, thiophene, malonitrile, tetracyanoethylene, and thianthrene, had activation energies of conduction below 0.9 eV. This finding is in marked contrast to the average of 1.36 eV for hydrocarbon and amino polymers. The authors [108] suggest that a different conduction mechanism is operative in the two groups of plasma polymers. A few sulphur-containing films were found in the highest conduction group. Polymers from thiophene and thianthrene exhibited conductivities higher than any hydrocarbon polymer. The presence of metal or transition elements did not result in special electrical properties.

References

1. C. L. Hammermesh and P. J. Dynes. *Polymer Letters, 13,* 663 (1975).
2. E. H. Cirlin and D. H. Kaelble. *J. Polym. Sci., Polym. Phys. Ed., 11,* 785 (1973).

3. D. H. Kaelble. *Physical Chemistry of Adhesion.* Wiley: New York (1971).

4. W. A. Zisman. "Surface Properties of Plastics," *Record of Chem. Progress,* 26(1), 13 (1965).

5. B. D. Washo. IBM Corp., Poughkeepsie, N.Y. (1977).

6. R. H. Hansen and H. Schonhorn. *Polymer Letters, 4,* 203 (1966).

7. L. A. Wall and R. B. Ingalls. *J. Chem. Phys., 35,* 370 (1961).

8. N. Saka, G. Y. Yee, and N. P. Suh. Presented in part at the Soc. of Plastics Engineers 35th Amer. Tech. Conf., Cleveland, 25 April 1977.

9. J. R. Hall, C. A. L. Westerdahl, A. T. Devine, and M. J. Bodnar. Tech. Rept. 3788, Picatinny Arsenal, Dover, N.J. (Jan. 1969).

10. B. R. Loy. *J. Polym. Sci., 44,* 341 (1960).

11. E. J. Lawton, J. S. Balwit, and R. S. Powell. *J. Chem. Phys., 33,* 395 (1960).

12. R. Salovey and W. E. Falconer. *J. Phys. Chem., 69,* 2345 (1965).

13. H. Schonhorn and R. H. Hansen. *J. Appl. Polym. Sci., 11,* 1461 (1967).

14. H. Schonhorn and L. H. Sharpe. *J. Polym. Sci., A-3,* 3087 (1965).

15. N. A. Melikhore, Z. P. Kosova, O. A. Kotovschchikova, and S. A. Reitlinger. *Sov. Plastics* (Eng. trans.), No. 4, 42 (1966).

16. N. A. de Bruyne. *Aircraft Engr., 18,* 53 (1939).

17. H. Yasuda. In *Plasma Chemistry of Polymers.* Dekker: New York (1976).

18. A. M. Trozzolo and F. H. Winslow. *Macromol., 1,* 98 (1968).

19. K. E. Bacon. *Newer Methods of Polymer Characterization.* Interscience: New York (1964).

20. J. R. Hall, C. A. L. Westerdahl, A. T. Devine, and M. J. Bodnar. *J. Appl. Polym. Sci., 13,* 2086 (1969).

21. A. Charlesby, *Atomic Radiation and Polymers.* Pergamon Press: New York (1960).

22. A. R. Schultz. *J. Chem. Phys., 29,* 200 (1958).

23. M. Hudis and L. E. Prescott. *J. Polym. Sci., B-10,* 179 (1972).

24. G. L. Weininger. *J. Phys. Chem., 65,* 941 (1961).

25. J. R. Hall, C. A. L. Westerdahl, M. J. Bodnar, and D. W. Levi. *J. Appl. Polym. Sci., 16,* 1465 (1972).

26. D. H. Reneker and L. H. Bolz. In *Plasma Chemistry of Polymers.* Dekker: New York (1976)

27. R. H. Hansen, J. V. Pascala, T. DeBenedictis, and P. M. Rentzepis. *J. Polym. Sci., A-3,* 2205 (1965).

28. R. B. Gillette, J. R. Hollahan, and G. L. Carlson. *J. Vac. Sci., Technol., 7,* 534 (1970).

29. Linus Pauling. *The Nature of the Chemical Bond,* 3d ed. Cornell University Press: Ithaca, N.Y. (1960).

30. J. R. Hall, C. A. L. Westerdahl, A. T. Devine, and M. J. Bodnar. *J. Appl. Polym. Sci., 13,* 2085 (1969).

31. J. R. Hollahan and G. L. Carlson. *J. Appl. Polym. Sci., 14,* 2499 (1970).

32. N. I. DePollis. Sandia Labs., Res. Rept. No. SC-RR-71-0920 (1972).

33. H. Schonhorn, F. W. Ryan, and R. H. Hansen. *J. Adh., 2,* 93 (1970).

34. J. R. Hollahan, B. B. Stafford, R. D. Fall, and S. T. Payne. *J. Appl. Polym. Sci., 13,* 807 (1969).

35. R. A. Auerbach. Int. Rept., Lord Corp. (9 Mar. 1978).

36. D. T. Olsen and A. T. Osteraas. *J. Appl. Polym. Sci., 13,* 1523 (1969).

37. D. D. Lidel. U.S. Pat. 4,072,769 (1978).

38. M. S. Akutin and L. A. Rodivilova. *Plast. Massy (6),* 2 (1960).

39. C. Simionescu, A. Nicolae, and D. Ferenez. *Cell. Chem. Technol., 3*(2), 165 (1969).

40. R. A. Auerbach and J. O'Connor. Internal memorandum, Lord Corp. (19 Oct. 1977).

41. R. E. Belin et al., *J. Text. Inst., 64,* 48 (1973).

42. M. M. Millard et al., *Text. Res. J., 42,* 307 (1972); *42,* 460 (1972).

43. M. M. Millard and A. E. Pavlath. *J. Macromol. Sci., Chem.,* A10(3), 579 (1976).

44. A. Pavlath and K. S. Les. In *Plasma Chemistry of Polymers.* Dekker: New York (1976).

45. D. C. Graham. *Chem. Revs., 41,* 441 (1947).

46. D. J. Shaw. *Introduction to Colloid and Surface Chemistry.* Butterworth: London (1970).

47. G. Err. *G. I. T. Fachz. Lab., 19,* 772 (1975).

48. O. Stern. *Z. Elektrochemie, 30,* 508 (1924).

49. L. Gouy. *J. Phys., 9,* 457 (1910).

50. D. L. Chapman. *Phil. Mag., 25,* 475 (1913).

51. P. N. Sawyer, *Surgery, 56,* 1020 (1964).

52. H. J. Jacobash. *Faserforsch. und Text. Technik, 20,* 191 (1969).

53. N. Saka, G. Y. Yee, and N. P. Suh. Soc. of Plastics Engineers 35th Ann. Tech. Conf., Cleveland, 25 April 1977.

54. P. H. Dundas and M. L. Thorpe. *Chem. Eng.,* 123 (1969).

55. D. A. I. Goring. *Pulp & Paper Mag. of Canada,* T372 (1967).

56. P. Blais, D. J. Carlsson, and D. M. Wiles. *J. Appl. Polym. Sci., 15,* 129 (1971).

57. J. J. Bickerman. *Science of Adhesive Joints.* Academic Press: New York (1971).

58. J. R. Huntsberger. In *Contact Angle, Wettability, and Adhesion.* Advances in Chemistry Ser., *No. 43.* Amer. Chem. Soc.: Washington, D.C., 180 (1964).

59. L. H. Sharpe and H. Schonhorn. In *Contact Angle, Wettability, and Adhesion.* Advances in Chemistry Ser., *No. 43.* Amer. Chem. Soc.: Washington, D.C., 189 (1964).

60. D. J. Carlsson and D. M. Wiles. *Can. J. Chem., 48,* 2387 (1970).

61. E. W. Garnish and C. G. Haskins. *Aspects of Adhesion, 5,* 259 (1969).

62. L. C. Jackson. *Adhesives Age,* p. 34 (Sept. 1978).

63. W. W. Balwanz. In *Surface Contamination,* Vol. 1, ed. K. L. Mittal. Plenum Press: New York (1979).

64. G. J. Kominiak and D. M. Mattox. Internatl. Conf. Metal Coatings, San Francisco, Cal. (8 April 1976).

65. P. J. Dynes and D. H. Kaelble. *J. Macromol. Sci., Chem., A10,* 535 (1976).

66. K. Bhasin, D. B. Jones, S. Sinharoy, and W. J. James. *Thin Solid Films, 45,* 195 (1977).

67. R. Jacobsen. *Thin Solid Films, 34,* 191 (1976).

68. L. W. Crane and C. L. Hammermesh. Presented in part at the Sympos. Adhes. Measurement of Films and Bulk Coatings, Amer. Soc. for Test. Matls., Philadelphia, Pa., Nov. 1976.

69. C. L. Hammermesh and L. W. Crane. *J. Appl. Polym. Sci., 22,* 2395 (1978).

70. R. G. Bosisio, M. R. Wertheimer, and C. F. Weissfloch. *J. Phys., E 6,* 628 (1973).

71. H. V. Boenig and D. R. Blenner. Lord Kinematics (1979).

72. L. Perrone. *Plastics Engineering,* p. 51 (1980).

73. M. C. Ross, U.S. Army, ARRADCOM, LCWSL, Tech. Rept. ARLCD–TR–77088 (1978).

74. S. L. DeGisi and C. H. Smith, *Adhesives Age,* p. 35 (July 1978).

75. C. A. L. Westerdahl, J. R. Hall, D. W. Levi, and M. J. Bodnar. Tech. Rept. 4279, Picatinny Arsenal, Dover, N.J. (1972).

76. R. H. Hansen, S. Hills, and H. Schonhorn. U.S. Pat. 3,462,335 (1969).

77. B. C. Taylor, W. V. Sharp, J. I. Wright, K. L. Ewing, and C. L. Wilson. *Trans. Amer. Soc. Artific. Int. Organs, 17,* 22 (1971).

78. H. V. Boenig and D. R. Blenner. Unpubl. data, Lord Kinematics (1979).

79. T. E. Nowlin and R. A. Martineau. U.S. Pat. 4,123,308 (1978).

80. M. S. Akutin and L. A. Rodivilova. *Plast. Massy (6),* 2 (1960).

81. E. L. Lawton. U.S. Pat. 3,853,657 (1974).

82. R. W. Baird and W. G. Baird, Jr. U.S. Pat. 3,870,610 (1975).

83. J. R. Hollahan and T. J. Wydeven, Jr. U.S. Pat. 3,847,652 (1974).

84. R. M. Kubacki. NASA, Tech. Briefs (winter 1978).

85. J. C. Fletcher, T. J. Wydeven, Jr., and J. R. Hollahan. U.S. Pat. 4,137,365 (1979).

86. R.C.A. British Pat. 1,445,546 (1974).

87. S. Aisenberg and R. Chabot. *J. Appl. Phys., 42,* 2953 (1971).

88. D. S. Whitmell and R. Williamson. *Thin Solid Films, 35,* 255 (1976).

89. L. Holland and S. M. Ojha. *Thin Solid Films, 48,* L 21 (1978).

90. S. Berg and L. P. Andersson. *Thin Solid Films, 58,* 117 (1979).

91. H. Biederman, S. M. Ojha, and L. Holland. *Thin Solid Films, 41,* 329 (1977).

92. R. A. Auerbach. U.S. Pat. 4,188,426 (1980).

93. D. A. Olsen and A. J. Osterhaas. *J. Appl. Polym. Sci., 13,* 1523 (1969).

94. F. Y. Chang, M. Shen, and A. T. Bell. *J. Appl. Polym. Sci., 17,* 2915 (1973).

95. K. R. Buck and V. K. Davar. *Brit. Polym. J., 2,* 238 (1970).

96. H. Yasuda and C. E. Lamaze. *J. Appl. Polym. Sci., 17,* 201 (1973).

97. J. R. Hollahan and T. J. Wydeven, Jr. *Science, 179,* 500 (1973).

98. S. Loeb and Sourirajan. In "Advan. Chem. Ser."; *Amer. Chem. Soc. Publ. 38,* 117 (1963).

99. A. F. Stancell and A. T. Spencer. *J. Appl. Polym. Sci., 16,* 1505 (1972).

100. B. Suryanarayanan, J. J. Carr, and K. G. Mayhan. *J. Appl. Polym. Sci., 18,* 309 (1974).

101. H. Yasuda and C. E. Lamaze. *J. Polym. Sci., A–2, 9,* 1537 (1971).

102. H. Yasuda and A. Schindler. In "Reverse Osmosis Membrane Research"; ed. H. K. Lonsdale and H. E. Podall; Plenum Press: New York (1972).

103. A. T. Bell, T. Wydeven, and C. C. Johnson. *J. Appl. Polymer Sci., 19* (1), 911 (1975).

104. H. McCain, U.S. Pat. 4,100,113 (1978).

105. J. C. Fletcher, M. Hudis, and T. Wydeven. U.S. Pat. 4,132,829 (1979).

106. P. J. Ozawa. IEEE, Trans. on Parts, Materials, and Packaging, PMP-5, No. 2, 112 (June 1969).

107. D. Ristow. *J. Materials Sci., 12,* 1411, 1977.

108. A. Bradley and J. P. Hammes. *J. Electrochem. Soc., 110*(1), 15 (Jan. 1963).

8. Additional Applications of Plasma Technology

1. Ion Implantation

A. Introduction

Implantation of ions in a metal surface is a technique that has shown a potential for producing vast improvements in wear, corrosion resistance, hardness, and fatigue lifetimes. It appears that improvements in the surfaces of optical elements and novel optical components may also be possible by this process. The application of this process to the improvement of the life of metal-cutting tools and its use in the semiconductor industry has been causing increasing interest. This chapter also describes how implantation of one metal into another can be used to explore the metallurgy of novel systems. It appears difficult to base generalizations and predictions upon existing knowledge. The atomistic phenomena that operate in ion-implantation effects are not well understood. Research efforts aimed at illuminating the mechanisms that underlie known beneficial effects of ion implantation in metals are presently under way and have been supported by the National Materials Advisory Board of the National Research Council [1].

Ion implantation is a process by which alloying elements are introduced into a host material by accelerating them to a high energy and permitting them to impact on the surface of the host. The impinging atoms penetrate into the substrate material to a depth of 0.01 to 1 micron, depending on the atomic number and energy of the atom, and form a thin, alloyed surface layer on the substrate. This process differs from others, such as electroplating, in that it does not produce a discrete coating; rather, it alters the chemical composition in the surface of the host material.

In more recent times, the electronic industry has made increasing use of this process as a means of doping semiconductors. Since the charge transferred to the surface determines the number of ions implanted, and the incident energy controls their depth distribution, ion implantation has markedly improved both control and reproduction of a number of semiconductor device-

260

processing operations. It is also noteworthy that ion-implantation processes generally do not require the high temperatures needed to induce ion diffusion into the base surfaces. Ion implantation is employed in the field of electronics to change the magnetic properties of substrates used for magnetic bubble devices, and the process has also attained industrial application by ion-nitriding steels [2]. It has been found that all steels that can be conventionally nitrided, as well as some alloys that cannot be processed at conventional nitriding temperatures, are suitable for ion nitriding. The advantages of this process are both energetic and environmental.

Ion-carbonitriding is the simultaneous implantation of nitrogen and carbon at temperatures of 1033 to 1123°K. The benefits from this plasma process include thicker hardened surface layers. This process requires quenching of the treated specimens, however, which frequently causes undesirable changes in part dimensions. It should be noted that nitriding in a plasma in the presence of a small amount, such as 0.5 atom % of carbon compounds, such as methane or acetylene, is generally not considered ion-carbonitriding, since it prevents the essential aspect of surface decarbonization.

Analogous to ion nitriding, ion carburizing also yields hard surfaces at 1123 to 1273°K. Nonferrous metals, such as Nb and W, can be carbonized in this process.

Borizing of metals in a glow discharge was first mentioned in 1973 [3] in experiments using a mixture of BCl_3 and 5% H_2. The temperature of the metal specimen was kept at 1053°K. After several hours FeB/Fe_2B layers were formed in the iron surface. The advantages of plasma borizing over conventional salt-bath borizing lie in the use of lower temperatures and the better control of layer formation.

The elementary processes taking place between such a plasma and the metal specimen have been described as follows [4].

The positive ions hitting the cathodic metal parts tend to remove the oxide and fatty layers, as well as other impurities (sputter cleaning).

Electrons are released that are needed for the maintenance of the gas discharge. The ions remove those atoms from the metal sample that can react with the gas atoms (metalloids).

The metal-metalloid compound can again reach the sample surface via diffusion and bouncing processes so that an integral layer is formed on the sample within a short period of time.

In this way a high concentration of metalloids, which is capable of diffusion, is formed on the surface, so that by decomposition of the compound due to thermodynamic instability, or under the influence of ion bombardment, diffusion can occur into the sample piece. In this way, new compounds with the matrix metal are formed by exceeding the solubility

limit; or, on the other hand, solvation can occur in the matrix, which leads generally to a crosslinking of the crystal structure in the metal.
A large part of the motional energy of the ions is converted into thermal energy, which leads to an increase of the temperature of the center, so that a diffusion process at reasonably large velocity occurs first.

Depending on the type of gas (metalloid carriers), various methods can be employed, as shown in table 8.1. In semiconductor applications, the work-piece into which ions are implanted is no larger than a wafer—an area of less than 100 cm². Areas of this magnitude are also encountered in ball bearings or lathe tools. Other applications, such as enhancement of corrosion resistance, may involve areas of many square meters. In semiconductor applications, high homogeneity is sought; the concentration of the implanted species should not vary by more than a small percentage over the surface of a wafer. The implantation of ions into semiconductors is generally patterned, that is, some areas of the substrate are covered by a mask that stops the incident ions before they enter the substrate. Implantation into semiconductors is usually followed by a high-temperature annealing. Other applications of ion implantation include modification of the properties of magnetic films for bubble devices [1] and integrated optics.

Table 8.1. Parameters in ion implantation

Process	Material	Plasma parameters
Nitriding	Steel	NH_3, N_2, N_2 + H_2 0.1–1.3 Pa 673–873°K max. 40 hr
	Cr-Steel-TiC	85% N_2 + 15% H_2 1 k Pa max. 1023°K 6 hr
	Ti-alloy	1073°K 20 hr
Borizing	Fe	95% BCl_3 + 5% H_2 650 Pa 1073°K
Carbonitriding	Steel	C-containing nitriding gas 983–1143°K 5 hr
Carburizing	Steel Nb, W	1123–1273°K

Source: B. B. Buecken, *Die Technik,* *33*(7), 395 (1978).

B. Properties

1. Hardness and Wear

English sources [5] reported recently the following improvements of metal surfaces by ion implantations.

Improvement in life by a factor of 4 is typical for steel components, such as paper and rubber cutting tools, as well as taps for plastics. A similar increase is found in cemented tungsten-carbide-wire drawing dies.

The benefits of ion implantation on wear may persist to a depth 10^3 times that of the implanted layer thickness. The implanted atoms are apparently transported into the metal as the tool wears.

Volumetric wear rates of a nitriding steel are improved by two orders of magnitude by nitrogen implantation at light loads.

Cutting tools used for machining high-strength materials for aircraft and weapon systems have been estimated to cost $900 million per year [5]. In view of the amount of machining done per year, improving wear life of cutting tools is of economic interest. Grinding, milling, turning, and broaching are major metal-removal operations. Cutting tools used for such work include end mills, taps, reamers, drills, and lathe tools. For ion implantation in cutting tools, carbon or nitrogen ions would be implanted in the edge surfaces of the tool. Depths of penetration of over 100 nm can be achieved with current techniques [1]. The use of ion implantation to improve tool life has the following advantages.

The low-temperature process results in negligible distortion.
Dimensional tolerances are unaffected.
Any element can be applied to any metallic substrate (limitations exist in glow discharges).

Two additional aspects make ion implantation to bearings attractive.

Dimensional changes do not exceed a few micro inches, which renders subsequent processing unnecessary.
The smooth transition between implanted surface layer and bulk reduces problems of surface-layer spalling.

It has recently been reported [6] that by injecting nitrogen into hardened steel ball bearings the wear rate was reduced by a factor of two.

Much more striking changes in hardness and wear rate were obtained by implanting a surface layer on softer metals. For example, boron injected into beryllium surfaces of precision gear components increased the surface hardness considerably.

2. Fatigue

Nitrogen implantation has been reported [7] to increase the fatigue life of a rotating carbon steel beam at loads of less than 90% of the yield stress and does not change it at stresses above the yield stress. An improvement by a factor of 8–10 in the fatigue lifetimes of nitrogen-implanted titanium stainless steel and maraging has also been observed [8]. These results appear to be consistent with the current understanding of the mechanism of fatigue failure. Fatigue cracks start at a surface, and there is a close connection between surface hardness and fatigue life. Compressive stresses due to the presence of additional implanted ions may also play a role in suppressing cracks.

3. Corrosion and Oxidation Resistance

Alloying has long been known to provide a route toward corrosion-resistant metals. Although the mechanism is not fully understood, it is known that incorporation of a galvanically more noble element will result in reduced corrosion. Elements such as chromium, nickel, titanium, and aluminum rely for their corrosion resistance on a tenacious surface oxide layer often referred to as passive film. Thus, alloying additions that produce improved passive films are used to obtain corrosion-resistant alloys.

Ion implantation is believed to increase corrosion resistance economically only in selected cases where bulk alloying is unsuccessful. This may apply:

If the desired alloying elements reduce a desirable property of the substrate, such as fatigue life or electrical conductivity.
If the costs are excessive.
If no appropriate bulk alloy is known.

Literature accounts indicate that ion implantation is assuming an increasing role in corrosion science and in the study of the atomistic mechanisms involved in corrosive reactions.

Most work to date has concentrated on ferrous alloys and titanium. The presence in iron of chromium results in the formation of a stable, tenacious oxide. When oxidation takes place by diffusion of oxygen through the surface layer, diffusion through the oxide layer is known to be slow.

4. Optical Properties

Ion implantation has been shown [9] to alter the indexes of refraction and optical absorption constants. The application of ion implantation to optics has been studied in two general areas, integrated optics (optical elements, such as wave guides and lenses) and individual components (such as filters, windows, and mirrors).

The field of integrated optics deals with the processing of light signals in insulating and semiconductor materials. In conventional electronic devices, the signals are carried by metal conductors. In integrated optics, the signals are guided within light pipes. Most important, integrated optics may provide the electrooptical interfaces that convert electrical to optical signals and vice versa and are needed to utilize optical fibers. Transmission of information in the form of light passing through an optical fiber is an important existing technology that provides, in addition to such advantages as security, considerable savings in weight in comparison with copper wires. Although optical and electrooptical devices are generally made by procedures such as thermal diffusion, ion exchange, and epitaxy, it appears that the advantages of ion implantation in the areas of control and reproducibility have attracted the industry in more recent times.

The Committee on Ion Implanation of the National Research Council pointed out [1] that the field of integrated optics is today at roughly the same state of development that semiconductors were 20 years ago. For some optical devices, the standard techniques suffice. Still, ''if the same level of complexity is to be achieved with integrated optics as now exists in semiconductor integrated circuitry, the traditional techniques will not be adequate. It is likely that ion implantation will have to play an important role in the next generation of integrated optical components.''

In other fields of optics, filters, mirrors, and windows are employed in a great variety of environments ranging from gun sights to satellite surveillance devices, including components for laser detection, communication, and sophisticated weapons. Tarnishing of metal mirrors has been successfully prevented by ion bombardment. Present ion-implantation work aims at protection of IR and UV transparent windows from atmospheres and micrometeorite deterioration. Ion implantation may also be an important process in the improvement of the optical properties of filters, mirrors, and windows. It is generally felt that more selective fibers can be made, since implantation provides better control of chemical doping.

C. Ion Nitriding

Bason [10] observed in 1928 that certain steels with a high nitrogen content exhibit a higher hardness than expected from their carbon content. This finding started early development on aluminum bearing steels but was soon extended to other elements such as chromium, vanadium, molybdenum, and titanium, all of which were found to form nitride layers. Conventionally, nitriding is carried out in an atmosphere of partially dissociated ammonia or in a cyanide-cyanate salt bath, at temperatures of 650-850°C.

Nitrided steels exhibit good corrosion and fatigue resistance and will maintain their hardness up to near-nitriding temperatures. Typical applications for nitrided parts include valves, gears, pump parts, piston pins, crankshafts, cylinder liners, shafting, bearings, and many other rolling or sliding contact parts.

Listed disadvantages of conventional nitriding processes have included [11]:

Storage of large quantities of ammonia.

Disposal of ammonia or salt-bath compounds as potential environmental hazards.

Difficulties in controlling growth of a "white layer," a thin, brittle surface layer, which tends to spall off under load. In general, its thickness must be limited to about 0.5 mil, which is said to be difficult with either gaseous or liquid nitriding.

Ion nitriding, on the other hand, is reported [11] to share the good features of conventional nitriding while offering additional advantages, such as:

Improved white layer control.

No problems with ammonia gas storage.

No problems in disposal and dissociation control.

Reduced floor-to-floor time.

Reduced distortion and growth.

Improved fatigue life.

Lower energy requirements.

In a so-called nitriding plasma the discharge properties are controlled by current density. The density of current increases until the glow covers the cathode completely; thereafter, any further increase must be accompanied by an increase in current density, and the discharge voltage then begins to rise. This represents the "abnormal" glow region and is characterized by a uniform current density on every point of the cathode. It is the region used for glow-discharge nitriding, uniform heating, and case development (figure 8.1). Transition from the normal to abnormal glow may occur at high currents; for while in laboratory experiments the electrode area may be only a few square centimeters, an industrial nitriding cathode may be several square meters in area and may require about 0.4 amps/sq ft.

Two other general discharge properties are to be considered, the hollow-cathode effect and sputtering. A hollow-cathode discharge occurs when opposed cathode surfaces are so close that the negative glows coalesce. Under such conditions, the current density can increase considerably and can result in localized nonuniformity of case depth. In sputtering, the cathode emits

Figure 8.1. Region of glow-discharge nitriding. *Source:* Adapted from K. S. Cho and C. O. Lee, *J. Engineering Matls. and Technol.,* forthcoming.

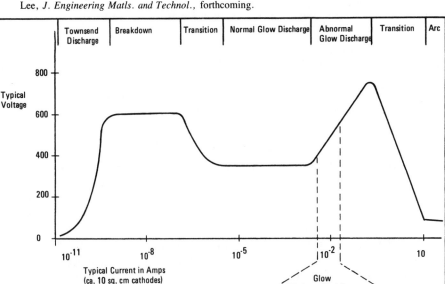

atoms from the surface in response to the ion impingement of the discharge. The rate of emission increases with ion energy, ion mass and, approximately, the number of electrons in the outermost "d" shell of the target atoms. The effective loss of cathode material at low pressures and large electrode distance is considerably reduced by back diffusion. The reactor is evacuated to a pressure of about 10^{-5} torr before a mixture of nitrogen and hydrogen is added. The d.c. discharge is ignited and pressure and temperature brought to the desired level. The furnace is electrically grounded and cool to the touch.

The positive nitrogen ions in the glow discharge are attracted toward the negatively connected work pieces. They impinge upon these surfaces, are thus occluded, and heat up the pieces to the required diffusion temperature. At this point, the nitrogen diffuses into steel in its atomic form. In thermal nitriding this is accomplished by dissociated ammonia, which is metastable at nitriding temperatures. Molecular nitrogen is essentially ineffective. In gaseous nitriding, ammonia is employed to ensure increased nitrogen solubility with α-iron by equilibration with surface ion nitrides and to provide a high nitrogen pressure to aid in nucleation of alloy nitrides. Considerably higher pressures would be needed in thermal nitriding if ammonia were replaced by molecular nitrogen.

The kinetics of ion nitriding are considered to be a complex phenomenon

Table 8.2. Changes in surface composition of stainless steel no. 304 during
ion nitriding

Condition	Cr	Fe	Ni
Untreated	23	64	13
Nitrided at 670°C	89	5	6
Nitrided at 475°C	23	71	6

Source: C. Braganza, S. Veprek, E. Wirz, H. Stüssi, and M. Textor, 4th
International Symposium on Plasma Chemistry, Zürich, Switzerland (Sept.
1979).

that cannot be described by a simple diffusion equation. Remarkable segregation processes are observed by nitriding stainless steel at temperatures between about 300° and 650°C. For stainless steel, nitriding proceeds at rates that are several orders of magnitude higher than those for titanium. For example, a weight increase of 1.8 mg/cm^{-2} corresponding to a ~25 μm thick nitride layer has been measured [12] after nitriding stainless steel no. 304 at 500°C and 0.8 torr for about 30 min. Table 8.2 shows the changes in weight percentage of the surface composition of stainless steel no. 304 during ion nitriding at 670°C and 475°C for a period of 1 hr, 43 min, respectively. At about 670°C, nitriding results in a strong enhancement of chromium in the nitrided layer, whereas iron-richer nitride is obtained at a lower temperature. These segregation processes are not limited only to the surface but extend into the bulk. Evidently, such effects will have significant impact on the mechanical and corrosion properties of steel nitrided at relatively low temperatures.

It has generally been observed that ion nitriding produces smaller, more evenly distributed carbonitrides that may aid in attaining more efficient gas equilibration with steel. It also appears that surface cleanliness, resulting from sputtering, and the removal of oxide layers by hydrogen, contribute to a more favorable absorption of nitrogen at the surface. The typical layer system of ion-nitrided steels consists of a maximum connective zone of ~ 30 μm that contains mainly iron nitride [4] and a 0.1 to 0.8 mm thick diffusion zone, the nitrogen content of which decreases toward the core.

The aforementioned white-layer structures observed in both gas nitriding and ion nitriding of ferrous metals are considered to be of importance. The white layer, also called compound layer, consists of a single carbonitride phase, the epsilon phase [13], which is uniform in thickness—about 15 μm—very compact and free from pores. If carbon is absent during nitriding, the white layer consists of a single nitride phase of Fe$_4$N and is referred to as "gamma prime phase." Its thickness is generally less than that of the epsilon phase. Even after long treatment times, it does not exceed a value of 10 μm. This layer exhibits very desirable ductile properties. A layer of this purity of a

Fe_4N (or γ-Fe_4N) layer can only be produced by ion nitriding, while gas nitriding yields a layer consisting of a heterogeneous mixture of γ-Fe_4N and ξ-Fe_3−2N with inherent stresses in the transitional regions between the different lattice structures. The improved mechanical properties of glow-discharge nitride surfaces are generally attributed to the monophase zone, since the γ-phase is considered to be more ductile than the ξ-phase.

A mixture of nitrogen and hydrogen has been found to give best results in ion nitriding. Very high nitrogen concentrations tend to yield excessive white layers. Replacement of hydrogen by other gases reduces the rate of case development. For most alloy steels, a mixture of 25% nitrogen and 75% hydrogen is often used for the heat-up periods. The nitrogen concentration is then increased to 50% during processing to increase the rate of case development. While the exact gas ratio is not critical, increasing the gas mixture above 50% results in a tendency to increase white-layer thickness.

The effect of ion nitriding temperature has a bearing on three important factors: (1) speed of case development; (2) maximum hardness attainable; (3) depth and composition of white layer. The general rule applies that the higher the nitriding temperature, the more rapidly the case will develop. Ion nitriding is considered unique in making it possible to develop a case at 315°C.

It can be seen that, above 650°C (1200°F), the case does not develop satisfactorily. The effect on hardness depth and case hardness of 32 CR MoV121D steel (0.32% C, 3% Cr, 1% Mo, 0.3% V) is shown in figure 8.2 for five different temperatures. Under these conditions, maximum case hard-

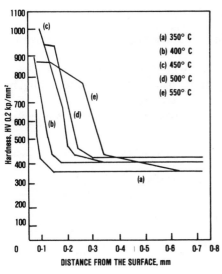

(a) 350° C
(b) 400° C
(c) 450° C
(d) 500° C
(e) 550° C

Figure 8.2. Hardness versus depth curves of alloy steel 32CrMoV1210 for 16 hours at 350°, 400°, 450°, 500°, and 550°C. *Source:* Adapted from A. M. Taylor, Advanced Vacuum Systems, Ayer, Mass. (1980).

ness is evidently developed at approximately 450°C. Development of the white layer is very time-dependent in gas nitriding, while ion nitriding produces this layer within the first few hours. Next to temperature, time is the most important variable in nitriding. The lower the temperature, the longer the time required to develop a given case depth. For most alloys, case development is somewhat more rapid in ion nitriding than in gas nitriding; for some alloys, such as AISI 4300 steels, the saving in time is outstanding. Rates of case development characteristics of different nitriding processes are shown in figure 8.3.

Figure 8.3. Rates of case development for AISI 4340 steel in different processes. *Source:* Adapted from A. M. Taylor, Advanced Vacuum Systems, Ayer, Mass. (1980).

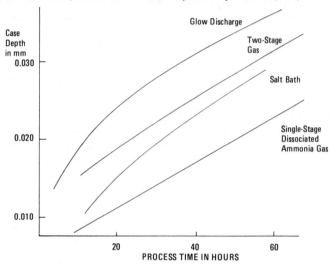

The variety of components that can be ion nitrided is almost unlimited and varies in size between very large gears to balls in ball-point pens. Highly complex shaped, parts, as well as simple ones, present no problems.

Successful ion nitriding of the following types of steels has been achieved [11]:

Al (containing low-alloy steels)
Cr (containing medium-carbon low-alloy steels)
Tool and die steels
High-speed-tool steel
Austenitic stainless steels
Ferritic and martensitic stainless steels
Constructional steels

Precipitation-hardening stainless steels
Carbon steel
Heat-resisting steels
Carbon-nitriding steels
Carburizing steels
Maraging steels

Iron nitriding has been achieved with the following cast irons:

Gray cast iron
Malleable iron
Ductile iron
Alloy irons

In the nonferrous field:

Iron (molybdenum alloys)
Titanium and titanium alloys
Tantalum
Molybdenum
Sintered alloys

Because of the wide range of conditions of temperature, pressure, and gas mixtures available, it is common to use a fixed gas pressure and mixture, varying the time and temperature. The standards initially recommended are shown below.

	During Warmup	Processing
Gas Mixture	75% Hydrogen 25% Nitrogen	50% Hydrogen 50% Nitrogen
Gas Pressure	2 torr	

Typical times and temperatures are shown in Table 8.3.

2. Effects of Carbon on Ion Nitriding

The effect of carbon content in the work piece as well as in the plasma on ion nitriding has recently been discussed [14]. Figure 8.4 demonstrates that the compound-layer thickness does not change with the carbon content in metal at 430 and 450°C but does markedly increase at temperatures above 470°C. At the same time, the diffusion-layer thickness is found to decrease in

Table 8.3. Times and temperatures for ion nitriding

Case depth required	Material	Time (hrs)	Temp °C
0.008–0.012″	nitralloys	10	525
	4140 steel	3	510
	4340 steel $\}$ Ni-CR-MO-V steel	12	510
0.015″ min.	nitralloys	30	525
	4140 steel	10	510
	4340 steel $\}$ Ni-CR-MO-V steel	12	510
0.025″ min.	nitralloys	—	
	4140 steel	30	510
	4340 steel $\}$ Ni-CR-MO-V steel	30	510

Source: B. B. Buecken, *Die Technik, 33*(7), 395 (1978).

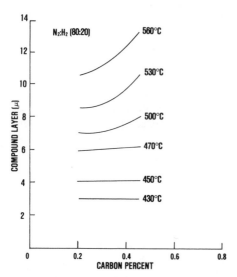

Figure 8.4. Relation between compound layer thickness and carbon content in steel (5 torr, 3 hrs., SCM3). *Source:* Adapted from K. S. Cho and C. O. Lee, Trans., *ASME, 102,* 222 (1979).

proportion to the carbon level in the metal. Similar general tendencies have been observed for other metals [15]. It has been suggested [16] that the compound layer is formed by a series of reactions:

$$Fe_3C \rightarrow Fe_3CFe_4N \rightarrow Fe_3CFe_3N \rightarrow Fe_{2\text{-}3}N$$

Accordingly, an increase in the carbon content in the metal would tend to increase the compound-layer thickness and decrease the diffusion-layer thickness, since carbon in metal tends to hinder diffusion of nitrogen. Figure 8.5

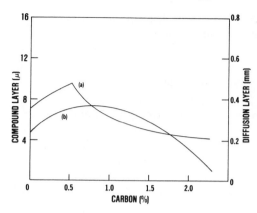

Figure 8.5. Effect of percent carbon in plasma (500°C, 5 torr, 3 hrs, SMC3) on (*a*) compound layer and (*b*) diffusion layer. *Source:* Adapted from K. S. Cho and C. O. Lee, trans., *ASME, 102,* 223 (1979).

shows the effect of percentage of carbon in plasma on both the compound and diffusion layers.

The maximum compound-layer thickness, that is, ϵ-monophase, is evidently obtained when 0.5 atm. % C is added, while for maximum diffusion-layer thickness, that is, γ + Fe_3C polyphase, 0.7 atm. % C is required in the plasma. It is also noted that maximum hardness is obtained, under the same experimental conditions, when 0.5 atm. % C is added to the plasma. In general, added carbon increases hardness at comparatively low temperatures because of increased compound-layer thickness and the formation of a fine grain size.

At this time, industrial applications of ion nitriding in Europe are close to 100,000 tons per year, involving the operation of over 70 units. In contrast, although no figures are published at this time, the United States is known to operate some 25 units of similar size.

Most ion nitriding presently has the following uses.

Extrusion cylinders for plastic processing equipment: Their service life is considerably longer than that of gas-nitrided counterparts, even though their case depth is much less (0.3 mm versus 0.5 mm). The difference is explained by the thin and tough monophased gamma prime-white layer and the ductile diffusion layer with suppressed segregations [13].

Gears: Ion nitriding considerably increases the wear resistance, gliding properties, and fatigue resistance of tooth flanks and tooth roots of gears. Because of layer properties and good dimensional stability, ion-nitrided gears do not require machining after treatment.

Motor vehicle construction: Small gears are ion nitrided on a production scale for several motor companies. Nitriding of truck crankshafts has become more popular because of the considerably higher bending and torsional fatigue strength obtained. A crankshaft for a Russian V-8 truck showed improved fatigue strength after ion nitriding when compared with gas nitriding. Shaft distortions were below 10 μm between neighboring bearings [13]. Gray cast iron discs for a rotary piston engine have been found to exhibit excellent wear and gliding properties, as well as high-dimensional stability when ion nitrided.

In the area of tool construction, the special layer qualities obtained by ion nitriding are of particular interest. All types of tools for hot and cold working, as well as for cutting, punching, and chipping are ion nitrided. Forging dies have been found to be particularly sensitive to improvement in hardness. The improvement stems from the very ductile and high wear-resistant layer produced in ion nitriding, although treatment time is only one-third that required in gas nitriding [13].

Cutting tools, such as drills, thread cutters, millers, and punches, are generally made from high-speed steels and hardened and annealed, followed by ion nitriding, with a reduced nitrogen potential to obtain complete suppression of the white layer for ductility reasons and a very thin diffusion layer of 20–30 μm.

In many machine tools, highly accurate spindles of different sizes are needed. They are ion nitrided to a hardness of dph 700 and to a case depth of 0.3 mm. Distortion must be below 20 μm and diameter growth less than 5 μm. For this application, "low-temperature" ion nitriding was chosen, that is, 20 hr at 450°C. Advantages of this lower temperature include high dimensional stability and higher hardness in the diffusion layer.

A. Ion Carbonitriding

Thicker layers of increased hardness are obtained if carbon and nitrogen are simultaneously introduced into the glow discharge at 1033° to 1123°K. In this case, however, it is necessary to quench the parts after plasma treatment, a step that frequently causes dimensional changes and therefore requires post-working.

Treatment of metals in a nitrogen plasma that contains only small amounts of carbon, such as 0.5% in the form of methane or acetylene, is not considered ion carbonitriding. While in carbon-free plasma the γ-Fe_4N nitride is formed in a thickness of up to 8 μm, ion carbonitriding suppresses this gamma phase in an Fe-C-N system. As an industrial process, this method has not gained much interest. It appears that many more data are needed to demonstrate the advantages of this technique.

B. Ion Carburizing

This process, which is the most common of the case-hardening processes, adds and diffuses nascent carbon into the surface of steel. Subsequent hardening by quenching from above 790°C produces a hard case in the high-carbon areas.

Carburizing is generally limited to low-carbon steels (below 0.25% C). The low-alloy steels do not harden deeply and result in lower core hardnesses. As alloying increases, the depth of hardness increases, and stronger and more uniform cores result. Because of the increased hardenability, alloy steels do not need to be quenched as severely, and distortion is less. Conventionally, carburizing is performed between 800° and 950°C. In this temperature range, nascent carbon is more soluble in steel because of the austenitic phase, and the diffusion rate is sufficient for economical use. By utilizing carburizing temperatures around 840°C, the diffusion rate is slowed down, allowing more time for raising the carbon content of shallow cases (up to 0.020 in. total) and to maintain control of case depth. Higher carburizing temperatures, generally around 930°C, are used for higher case depths.

Recent studies on vacuum carburizing have shown that carburizing of steel at a high temperature of 1048°C in a glow-discharge methane plasma greatly reduces carburizing time [17]. Plasma carburizing is analogous to ion nitriding. The specimen or work piece is made by the cathode in a d.c. electrical circuit. A counterelectrode, the anode, is placed near the work piece, and a glow discharge is established at a suitable pressure in the range of 1–20 torr. Under these conditions, approximately 400 V will sustain the plasma in a methane atmosphere, in which the methane is dissociated and a high carbon potential established.

Hardness profiles for some carburized cases on AISI 1018 steel obtained by three methods are shown in figure 8.6. The lower hardness region for vacuum carburizing applies to treatment without plasma at 20 torr, while the higher hardness region was obtained at a carburizing pressure of 300–375 torr.

It has been pointed out [17] that sufficient carbon can be added in only 10 min from a methane plasma to provide an effective case depth of approximately 1 mm when diffused for an additional 30 min in vacuum. With such a short carburizing time and low reactor pressure of 20 torr, plasma carburizing, if adopted, would provide significant savings in the amount of methane or natural gas consumed. The increased carburizing rates achieved in a plasma may be attributed to an increased rate of infusion of carbon into the surface, which is brought about by the high carbon potential. The rate of diffusion of carbon into the inner surface is believed to be governed by the same mechanism as that observed in conventional processes. While the weight percent of carbon on the surface may be 1.6%, it is approximately 0.5% and 0.2% at 0.5 mm and 1.0 mm depth, respectively.

Plasma carburizing points to an additional mechanism that may affect the rate of infusion of carbon into the surface. Since the dissociation of methane generates hydrogen, surface oxides are subject to reduction. Thus, a potential

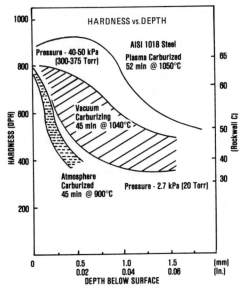

Figure 8.6. Hardness profiles of AISI 1018 steel carburized by three processes. *Source:* Adapted from W. L. Grube and J. G. Gay, *Metallurg. Trans.*, *9A*, 1421 (1978). Copyright by American Society for Metals and The Metallurgical Society of AIME, 1978.

barrier to introduction of carbon into the surface is removed. Plasma carburizing also reportedly [17] tends to diminish the generally observed differences in carburizing rates between various steels.

In addition to an increased carburizing rate, other concurrent advantages of plasma carburizing include uniform case depth and elimination of sooting. Since the methane is dissociated principally by the plasma at the work-piece surface, a high carbon potential exists only in this region and therefore diminishes the tendency of free carbon to form in areas remote from the work piece.

C. Ion Borizing

Gifford [18] reported on experiments involving the borizing of steel in a glow discharge in which he used BCl_3 and 5% H_2 at 6.5×10^2 Pa. The current density of the sample was adjusted in such a manner that it resulted in a temperature of $1,053°K$. After a few hours, layers of FeB/Fe_2B are formed. The advantage of plasma borizing, as compared with the conventional salt-bath borizing, lies in the application of lower temperatures as well as in the more advantageously directed layer formation. Because of its toxicity, however, this method is not yet used in the industry.

More recently, it has been shown [19] that ion implantation to produce a buried layer of boron in beryllium results in boron concentrations greater than those attainable with conventional thermal treatments.

3. Microcircuit Fabrication

A. Description

A typical conventional process is compared with plasma processes in figure 8.7. Plasma cleaning is a dry process that requires elevated temperatures, 40–60°C and oxygen or oxygen-containing substances. Low-temperature oxidation enables the volatilization of organic substances without creating undesirable oxides on the surface.

Plasma chrome photomask etching—dry etching of chrome metal from glass substances—is considered a safe process. This is in contrast to the more difficult wet process that is reported [20] to be hard to control and difficult to

Figure 8.7. Conventional process versus plasma processes. *Source:* Adapted from LFE Corp., Process Control Bull. No. 8200-SBl, April 1976.

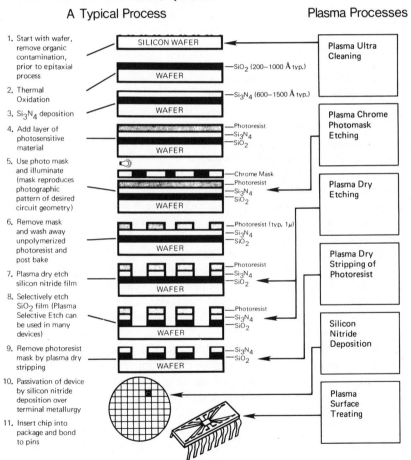

A Typical Process   Plasma Processes

1. Start with wafer, remove organic contamination, prior to epitaxial process
2. Thermal Oxidation
3. Si_3N_4 deposition
4. Add layer of photosensitive material
5. Use photo mask and illuminate (mask reproduces photographic pattern of desired circuit geometry)
6. Remove mask and wash away unpolymerized photoresist and post bake
7. Plasma dry etch silicon nitride film
8. Selectively etch SiO_2 film (Plasma Selective Etch can be used in many devices)
9. Remove photoresist mask by plasma dry stripping
10. Passivation of device by silicon nitride deposition over terminal metallurgy
11. Insert chip into package and bond to pins

SILICON WAFER

WAFER — SiO_2 (200–1000 Å typ.)

WAFER — Si_3N_4 (600–1500 Å typ.)

WAFER — Photoresist — Si_3N_4 — SiO_2

WAFER — Chrome Mask — Photoresist — Si_3N_4 — SiO_2

WAFER — Photoresist (typ. 1μ) — Si_3N_4 — SiO_2

WAFER — Photoresist — Si_3N_4 — SiO_2

WAFER — Photoresist — Si_3N_4 — SiO_2

WAFER — Si_3N_4 — SiO_2

Plasma Ultra Cleaning

Plasma Chrome Photomask Etching

Plasma Dry Etching

Plasma Dry Stripping of Photoresist

Silicon Nitride Deposition

Plasma Surface Treating

reproduce. As many as 15 photomask plates (5 in. $\times$ 5 in.) can be processed at one time. All gases and effluents are said to be nontoxic, nonflammable, and noncorrosive (Department of Transportation green label classification) and require no special plumbing, clean-hood facilities, or handling procedures.

Plasma dry etching of IC wafers is more often used in place of wet chemical etching. Some applications are [20]:

Rapid etching of Si_3N_4 using standard photoresist (thermoset) lithography.
Etching of SiO_2 and other dielectrics down to aluminum metallization layers.
Etching high-resolution openings (less than one micron), limited to photo-lithography.
Etching silicone prior to metallization of contact points for the enhancement of adhesion and reduction of ohmic contact resistance.
Smooth etching of doped and undoped polycrystalline silicone.
Etching of refractory films used in double-level metallurgy structures.

Plasma dry etching does not adversely affect GaAs, alumina, indium anti-monide, garnet, nickel, and aluminum.

B. Plasma Photoresist Stripping

A commonly used technique in transposing a pattern to a metal or insulator film is by photolithography deploying a photosensitive polymer generally referred to as photoresist. The image is reproduced onto the photoresist by a mask pattern followed by exposure of the photoresist to UV light. In this way the pattern is reproduced from the mask to the wafer. After the exposed material has been removed by plasma (or wet chemical etching), the photo-resist mask is then taken off.

This final removal of the crosslinked polymer is a generally tedious process and depends on the nature of the photoresist film, its thickness, and the underlying substrate. The wet process involves hot chlorinated hydrocarbons, followed by oxidizing agents such as hot H_2SO_4 and swabbing or brushing operations. The jet action of spray rinse is generally not sufficient, thus making a second rinse necessary.

Plasma stripping, on the other hand, is said to have been highly successful [21]. In this process, the highly reactive media convert organic photoresist to low-molecular-weight volatiles such as CO_2 and H_2O. Plasma photoresist stripping is preferred in the semiconductor industry for many reasons:

Speed: Typical rate is 50 to 100 3 in. silicone wafers in 10 to 20 min.
Costs: Expensive wet chemicals are eliminated and substantial cost savings are claimed.

Yield: Elimination of all organic films prior to the diffusion processing of the wafers as well as reduction of process steps.

Safety: Dry-plasma processing eliminates hazards to operating personnel, since no solvents are involved.

A mixture of oxygen gas and organo-halide vapor is often preferred. Apart from decomposing and volatilizing organic photoresist material, the plasma effects simultaneously a marked reduction in the quantity of inorganic contamination, especially that of tin, iron, and magnesium embedded in photoresist material. It also prevents exposed areas of semiconductor substrates from being covered by an undesirable oxide film, thus reducing ohmic contact resistance prior to metallization steps.

In most cases, the photoresist consists of a crosslinked polyisoprene chain structure as in Figure 8.8: The photoresist may also contain structural units

$$
\begin{array}{c}
\text{---C---C} {=} \text{C---C---} \\
\mid \\
\text{C}
\end{array}
\qquad \textit{Figure 8.8}
$$

such as in Figure 8.9. The stripping (oxidizing) plasma, when used with a photoresist of the structure in figure 8.8, reportedly [21] requires longer plasma-stripping time than that for structures shown in figures 8.9 and 8.10. Thus it seems that pendant unsaturation is more readily attacked in such a plasma than is the unsaturation in the polymer backbone.

$$
\begin{array}{c}
\text{---C---C---} \\
\mid \\
\text{C---C} \\
\parallel \\
\text{C}
\end{array}
\qquad \textit{Figure 8.9}
$$

$$
\begin{array}{c}
\text{C} \\
\mid \\
\text{---C---C---} \\
\mid \\
\text{C} {=} \text{C}
\end{array}
\qquad \textit{Figure 8.10.}
$$

The possibility of cyclization of structures, as shown in figures 8.9 and 8.10, to form stable hexagonal rings (figure 8.11) with the simultaneous elimination of double bonds has been reported [21] to increase oxidation resistance and reduce fragmentation.

Active oxygen in plasma discharge will attack unsaturated bonds; this is due to the electrophilic character of atomic oxygen. It has also been observed

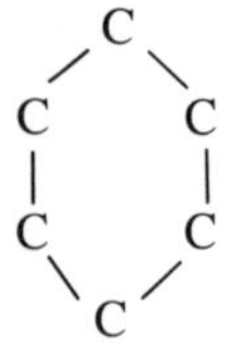

C
/ \
C   C
| |
C   C
\   /
C

Figure 8.11.

that plasma stripping (oxidation) of polymeric material is not affected by the crystalline content in the polymer. It appears that the active species in the oxidizing plasma have sufficient energy to attack those parts of the polymer chain exposed at the surface without regard to amorphous or crystalline areas, although the rates may differ.

A number of species have been identified in the oxygen plasma (chapter 3). The "low-temperature" oxygen plasma consists of 10–60% atomic oxygen, depending on reactor conditions. The concentrations of ions (O^+, O^-) and electrons are lower by four to five orders of magnitude.

The initial attack of oxygen atoms on photoresist is believed to be the abstraction of hydrogen to form carbon and oxygen-containing free radicals and water. Subsequent reactions of O atoms on the free radicals yield CO, CO_2, OH, and hydrogen atoms, possibly through reactions such as photoresist, that is,

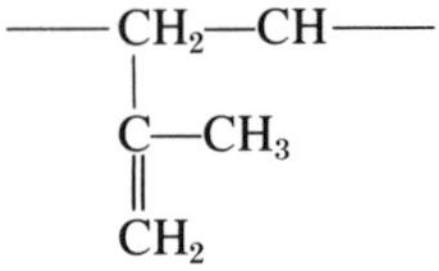

$$-CH_2-CH-$$
$$|$$
$$C-CH_3$$
$$||$$
$$CH_2$$

1. $CH_3 + O^* \rightarrow H_2CO + H$
2. $HCO + O^* \rightarrow OH + CO$
3. $HCO + O^* \rightarrow H + CO_2$
4. $OH + O^* \rightarrow O_2 + H$
5. $O^* + CO + M \rightarrow CO_2 + M$
6. $OH + H_2CO \rightarrow H_2O + HCO$
7. $O^* + H_2CO \rightarrow CO + H_2O$

Reactions 6 and 7 have a lower probability of occurrence. The yield of CO_2 in this system is only 3% to 6% because of the relatively slow triple collision reaction $CO + O + M \rightarrow CO_2 + M$ (M = third body).

This reaction emits the blue glow of visible radiation characteristic of electronically excited CO_2 molecules. The yield of CO_2 is independent of pressure, temperature, and the presence of hydrogen atoms indicating a slow reaction rate between CO and O.

The rate of removal of organic polymeric material can often be increased in an oxygen plasma where small amounts of gaseous species, such as water vapor and H_2, are added. It is believed that such catalytic activity is due, at least in part, to wall effects, that is, by rendering the inner walls of the reactor more passive toward atom recombination processes. Considerably increased O atom generation has also been observed by addition of N_2, NO, and N_2O. No measurable effects have been found by the addition of He, Ar, or CO_2.

Pure O_2 yields 0.9% of atomic fraction

O_2 + N_2 (0.01–0.05%) yield 90–100 O atoms/N_2 molecules
O_2 + H_2 (0.01–0.05%) yield 150–200 O atoms/H_2 molecules
O_2 + NO (0.01%) yield 45 O atoms/NO molecule
O_2 + N_2O (0.05%) yield 100 O atoms/N_2O molecule

Because of these large catalytic effects coupled with relatively short residence times within the discharge, surface effects are eliminated as a sole explanation of the phenomena. The high efficiency of N_2 and H_2 suggests chainlike sequences consisting of very fast ion-molecule or ion-electron reactions:

$$NO^+ + e \rightarrow N + O$$
$$N + O_2 \rightarrow NO^+ + O$$

The key features of plasma stripping have been summarized as follows [21]:

It is noncorrosive to underlying substrates.
It is free from ionic impurities.
The cost is low compared with wet process.
Process steps are reduced.
Chemical storage and disposal are eliminated.
Safety hazards for operating personnel are absent.
It is reproducible.
It requires less floor space.

C. Silicon Nitride Deposition

Semiconductor manufacturers have been protecting silicon wafers for years by coating them with a final silicon dioxide layer. This process does not seal the wafer hermetically, however. It uses a material, such as silicon nitride, requiring such a high temperature that the metals are often adversely affected.

More recently, a silicon-nitride-plasma process has been introduced to the industry that requires temperatures of only about 300°C and hermetically seals plastic packages for more expensive units. Silicon nitride Si_3N_4 is plasma-deposited utilizing diluted silane and nitrogen reactant gases; nitrogen gas is dissociated and reacts via a complex mechanism with silane to form a silicon nitride film on a heated surface. This plasma-deposition process can be used in applications such as:

Encapsulation of terminal metallurgy in low-cost plastic packages.
Metal-stop coverage required in double-level metallurgy.
Final nitridation of gate oxide.
Formation of thin-film capacitors.

4. Membranes

One of the important properties of a plasma-deposited thin film is its integrity. It appears that there exists no technology other than plasma deposition for obtaining thin, pinhole-free films that are tightly bonded to the substrate surface. Such films are potentially useful as membranes. Many of their properties have been discussed in chapter 7.

Some of the earliest applications for plasma films were as protective coatings. As protective liners inside beverage containers they appeared to be economically competitive with the existing alternatives but lacked suitable technology. Other applications were as coatings on metals for corrosion resistance and oxidation protection [22]. Emphasis appears to have been shifting toward the creation of selectively permeable films. It has been reported [23] that permeation through nitrocellulose films is reduced by exposing them to toluene discharges. This increased resistance to permeation was attributed to surface crosslinking of the nitrocellulose rather than to a toluene polymer film. Plasma polymers from 16 different monomers on filter paper were recently tested as barriers for water vapor permeation [24], with varying degrees of success. It was also observed that while dry tensile strength of the treated filter paper decreased, wet tensile strength increased as much as 34 times. Yasuda and Liepins [25] extended this study to Kraft fiber mats, one with and one without an intermediate collodian coating but both employing plasma deposition of styrene. Furthermore, reverse-osmosis membranes, prepared by plasma polymerization of several nitrogen-containing compounds, displayed very high salt rejection and high water fluxes (chapter 7). Unexpectedly, these properties improved with use for time periods of up to 80 days.

Another recent report [26] disclosed a H_2/CH_4 permselectivity increase of 297 across a poly(phenylene oxide) membrane with a plasma-deposited film of cyanogen bromide. The investigators concluded from their data that the coatings were effective at discriminating among nonpolar gases on the basis of molecular size. Others [26] reported reduced leaching of material from the bulk membrane and lowered gas permeability for plasma-deposited films. Decreased permeability of various gases through plasma-polymerized polypropylene films has also been observed [26].

5. Biomedical Applications

More recently, Hollahan and colleagues [27] reviewed the potential biomedical uses of plasma-deposited polymers and summarized their work to date. The basis of this work was to exploit the compatibility of plasma polymers with biological tissue, which rivals that of the best synthetic materials.

The earliest application in this field took advantage of the high reactivity of plasma gases to alter the surface of conventional polymers. Ammonia and hydrogen-nitrogen plasmas were employed to introduce amino groups to which heparin was complexed. Blood coagulation on the heparinized surface was found to be retarded but not prevented. In another investigation [28] plasma-polymerized ethylene film was used as a protective coating to reduce leaching of the base polymer by pseudoextracellular fluid (PECF).

6. Optical Applications

More recent applications of plasma-polymerized films have taken advantage of the thin film's optical clarity, smoothness, absence of pinholes, moisture resistance, and ease of deposition. Such films were also applied to conventional optics as combination antireflection, moisture-resistant, and abrasion-resistant coatings [29].

Integrated optics is a technology that has developed since 1968. It incorporates entire optical systems into a single device similar to an integrated electronic circuit. The state of the art of integrated optics has been summarized recently [30]. Many of the criteria for passive connections between components of integrated optical circuits are met by plasma-deposited film. They include the need for the connections to be made by thin, extremely smooth materials possessing a predictable and stable refractive index of a value consistent with the components for satisfactory long-term performance. Plasma-polymerized organo-silicon films have been reviewed for such use [31]. For their best films, the light loss was reported to be 0.04 dB/cm, a tenfold improvement over the best conventional films. The refractive index did not vary more than 1 part out of 1000 over a period of 6 months. More recently, the formation of such thin films for use as light-guiding interconnectors has been described [32] and extended to part of three-dimensional integrated optical circuits [33].

7. Review of Surface Treatments

Plasma surface treatments can be classified into the following groups:

Recombination
Oxidation and oxidative degradation
Peroxide formation
Interfacial recombinations
Grafting
Halogenation
Nitriding, borizing, wiring.

In addition, oxidation rates for 36 polymers in an oxygen glow discharge led to the suggestion that the polymer might be suitable for use in the upper atmosphere [34].

Plasma treatment of polymer surfaces as a means of improving the adhesion characteristics of polymers, inherently poor, has been discussed in chapter 7, section 2, including tenfold or greater adhesive joint strength by the use of the CASING technique [35]. Discharge-induced autohesion of polyethylene was ascribed to the formation of electrets [36]. Lawton [37] investigated adhesion of latices to poly (ethylene terephthalate) fibers following the fibers' exposure to low-temperature plasmas. For this system the relative adhesion by various gas plasmas was

$$NH_3 > N_2 > Ar \simeq He > air > O_2$$

Exposure time, power level, pressure, and flow rate were said to have no influence on adhesion within the limits tested.

8. Plasma Cleaning of Metal Surfaces

O'Kane and Mittal [38] demonstrated that mild plasma cleaning of metal surfaces is effective in removing organic contaminants. The effect of contaminant removal on adhesion has been discussed in greater detail in chapter 7, section 2F, of this book. Surface wettability has also been shown [39] to be related to the amount of contaminants present on the surface of metals and other materials.

Transition metal oxides are among the most sensitive surfaces for detecting contaminants, since they exhibit the most pronounced change in wettability. Erb [40] suggested earlier that a wettable oxide surface exists on rhodium after the usual organic contamination is removed. Elimination of this oxide layer resulted in a nonwettable surface. Wettability of other metals, such as chromium and stainless steel, has also been explained in terms of surface oxide layers. Fowkes [41] demonstrated earlier the inherent hydrophobic nature of oxide-free metal surfaces by considering the dispersion forces involved in surface wetting. It has also been shown [42], however, that the surface of pure gold is hydrophilic in nature.

O'Kane and Mittal [38] conducted an interesting experiment by measuring the rate of adsorption of hydrophobics on the surface of Fe-Co when, after plasma cleaning, the surfaces were exposed to laboratory air. They measured an increase in contact angle from approximately 5° to about 40° within 3 hr. These investigators also demonstrated that plasma cleaning of metal surfaces with argon or helium-oxygen is more effective than cleaning with organic solvents. The plasma cleaning was mild enough to avoid damage to the magnetic properties of a Fe-Co film.

Chrome photomasks fall into the same category as thin-film resistors in that they are easily oxidized. Therefore, the use of pure oxygen is not acceptable, and other gas blends must be considered. Plasma cleaning of photomasks has been used to clean residual resist transferred to the mask during contact printing operations with silicon wafers. In this application a very short cycle time is required (5–10 min). In addition, a very thin oxide may be formed on the mask that will harden the surface. Plasma cleaning is also used in the fabrication of the mask itself.

9. Plasma Etching

A. Etching of Polymer Surfaces

Selective etching of polymer surfaces has found interest primarily in areas such as increasing their roughness, changing the chemical constitution, degrading or dissolving low-molecular-weight materials that migrate to the surface, relieving residual surface stresses, studying the morphology of surfaces, and examining the internal structure of fibers.

The etching of polymers by plasma is a technique frequently used in the preparation of samples for electron microscopy [43,44]. Because of the differing resistance of morphological structures to etching, morphological detail can be exposed and in this way made accessible to examination by electron microscopy, which may involve the making of a surface replica, or direct examination by scanning electron microscopy (SEM).

There have been reports in the literature regarding the etching of oriented semicrystalline polymers by the plasma technique. They involved correlation of morphological features (markings 50 Å to 200 Å in diameter, oriented perpendicular to the draw axis) with the long period as determined by X-ray scattering (SAXS) [45, 46, 47]. Other investigations involved the observation of much larger morphological features, about 1000 A, again oriented perpendicular to the draw direction. It was inferred from these observations that such units form an integral part of the microstructure of oriented semicrystalline polymers [48,49].

More recently, Garton and colleagues demonstrated [50] that these larger structures can be artifacts of the activated gas-etch process resulting from restructuring of partially etched material. Accordingly, an etching agent should satisfy various criteria.

The etching action should differentiate between morphological features by, for instance, differences in density or permeability.

The decomposition products of the etched material should be completely removed.

The etching should not cause excessive restructuring of the surface.

Plasma appears to satisfy the first condition in that the amorphous component of the semicrystalline material is removed first [48,49,50]. The selectivity of plasma etching may be largely due to density differences but less so for fibrillar structures, in which density differences are smaller. The second criterion calls for selective plasma conditions under which the decomposition products are cleanly removed, that is, slow etching rates and rapid removal of gaseous products by a high pumping rate followed by a brief solvent wash. As to the third criterion, artifacts caused by restructuring have been adequately described [48]. Garton and colleagues [50] state that restructuring is only prevented by the presence of tie molecules in the amorphous phase. Plasma etching will fracture these tie molecules and induce extensive restructuring unless the partially degraded material is rapidly removed. If not, relaxation of the oriented polymer will occur, resulting in longitudinal shrinkage and, as a consequence, the appearance of corrugated structure, occasionally referred to as "sea-shore" structure, oriented perpendicular to the filament axis. The restructured surface is removable, however, by chemical etchants, such as a brief chromic acid treatment, thus exposing the sample interior to microscopic examination.

The trend in the semiconductor integrated-circuit technology is toward greater density of components on a device chip. This density requires smaller and narrower lines, more precise definitions for these compounds, and better controls. In the fabrication of electronic devices such as field-effect transistors, a base semiconductive material of a given conductivity type is used. In order to produce distinct regions of opposite conductivity type within the semiconductor material, impurities are diffused into its surface.

For example, if an N-type silicon crystal is employed as the semiconductor material, distinct regions characterized as P-type may be formed by diffusing boron into the silicon. Between any two such exactingly spaced P-type regions lies a zone characterized as N-type. Overlaying this zone is a thin film of dielectric material on which is deposited a metallic electrode. Connecting any two of the P-type regions in an electrical circuit and connecting the metallic electrode to a power source completes a typical field-effect transistor.

The insulating film is of utmost significance in the operation of field-effect transistors. The film should be thin with a high dielectric constant, since the thinner the film and the higher the dielectric constant, the greater is the induced carrier concentration in the N-type zone. It appears that films made from *p*-xylylene are the preferred material, since they have satisfactory dielectric constants and breakdown voltages at thicknesses between 100 Å and 10,000 Å. Furthermore, both the dielectric constant and dissipation factors are relatively constant over a wide frequency range. It is also important that in the coating process the substrate is not adversely affected.

Since the film will not cover the entire substrate, the difficulty has been to produce devices with only desired areas covered by the film. In a U.S. patent disclosure [51], this problem has been solved by plasma-depositing a film of *p*-xylylene polymer onto the entire substrate. A perforated metal mask is then placed over the film and the exposed areas of the film etched in a nitrogen glow discharge. The patent claims that an extremely fine etch delineation can be obtained in this manner. Plasma-etching techniques have also been employed to examine the internal structure of synthetic fibers [52]. More recently, Blakey and Alfy [53] plasma-etched polyethylene terephthalate fibers in a r.f. glow discharge at 50 W power. Oxygen was used as the ionized gas for masking the surface of the fibers. After an etching time ranging from 5 to 30 min, the specimens were examined in a scanning electron microscope.

Three types of internal flaws were observed by this method. The longest voids had their long axes parallel to the fiber axis and displayed regularly spaced bridges. They are believed to be caused by air or gas bubbles trapped in the molten polymer during extrusion. The investigators also observed short and intermediate-sized voids.

B. Etching of Inorganic Surfaces

1. Introduction

During the last decade, the complexity of integrated circuits (ICs) has greatly increased, and chips of area in excess of 5 mm^2 are now being mass produced. For a maximum number of functions or memory cells per wafer and increase of device performance, the fabrication of device patterns with smaller dimensions has become the continuing demand of device engineers.

Plasma-etching of silicon nitride was one of the first processes to be employed in production. A typical etch cycle was recently reported [54] to

Table 8.4. Comparison of wet and dry nitride etching

Wet etch	Dry etch
Deposit oxide	Develop resist pattern
Develop resist pattern	Etch nitride (CF_4-O_2)
Etch oxide (buffered HF)	Remove resist
Remove resist	
Etch nitride (hot phosphoric acid)	
Remove oxide	

Source: R. C. Booth and C. J. Heslop, *Thin Solid Films, 65,* 111 (1980).

consist of a prewarm (30 min in O_2 plasma at 3 torr and 200 W), loading the wafers and the backing discs, 2 min in an O_2 plasma to remove resist residues and for wafer etching in CF_4-O_2. Table 8.4 illustrates the advantages of this process over wet etching.

2. Aluminum

The wafers are placed on the lower, earthed aluminum electrode, which rotates to ensure uniform etching of a batch. The etching gas is CCl_4 vapor, which is excited by a 380 kHz r.f. voltage applied to the top aluminum electrode.

The etching process involves reaction of chlorine from CCl_4 with aluminum to form Al_2Cl_3. A liquid-nitrogen cold trap condenses CCl_4 vapor and other volatile reaction products. An etched depth of about 1 μm is obtained within 3 min. The aluminum etches faster at the periphery of the wafer than at the center. The etching is believed to be slowed at the center by a surface layer of aluminum oxide.

3. Silicon Dioxide

Using the same parallel plate system as in aluminum etching, Booth and Heslop [54] obtained satisfactory etching by using C_3F_8 as the etching gas. Peak selectivity occurred at a pressure of 0.15 torr and 380 kHz r.f. power.

A recent U.S. patent [55] describes the use of UV radiation during plasma etching of silicon dioxide. Because of the opacity of air and various glass, quartz, and crystalline window materials below wavelength of about 2000 Å, the inventors operated with a UV region of 600–1000 Å. The wavelength range corresponds to a photon energy range of 60 keV to 6 eV. The ionization potentials of most organic compounds including fluorocarbons lie in the energy range 8–13 eV, so that they will form ion species under such UV irradiation according to the following reactions:

$$CF_4 \rightarrow C\cdot + 4\ F\cdot \xrightarrow{\ O_2\ } CO_2 + 4F\cdot$$
$$4F\cdot + SI \rightarrow SiF_4$$
$$4F\cdot + SiO_2 \rightarrow SiF_4 + O_2$$

4. Silicon Nitride

Hampy [56] reported recently on the deposition and etching of silicon nitride films employing a r.f. plasma. Si_3N_4 films of good quality for semiconductor application were obtained in thicknesses ranging from 300 to 10,000 Å, with refractive indexes between 2.0 and 2.1 at deposition rates of 250–300 Å/min and about 300°C deposition temperature. For plasma etching of Si, SiO_2, and Si_3N_4, $CF_4 + O_2$ is generally used. The fluorine species in the plasma react with silicon nitride to form SiF_4 [57]:

$$CF_4 \rightarrow C \cdot + 4F \cdot$$
$$\rightarrow :CF_2 + 2F \cdot$$
$$12\ F \cdot + Si_3N_4 \rightarrow 3SiF_4 + 2\ N_2$$

The intended use for the Si_3N_4 film in semiconductor devices is as a protective coating over the SiO_2 layer. Since the wafers at this point in processing will have gone through several chemical and photoresist steps, Hampy [56] found it advantageous to remove any residual organic matter before etching by a pre-etch step in O_2 plasma prior to deposition. He reported the following optimum etching conditions after Si_3N_4 deposition: ambient temperatures, 0.05–0.07 torr pressure, 250–300 watts, 92 sccm CF_4 flow and 8 sccm O_2 flow.

Under these conditions, the etch rates were linearly dependent upon the power setting, ranging from about 420 A/min at 250 watts to about 950 A/min at 500 watts.

5. LiNbO₃

In integrated optics (IO), $LiNbO_3$ is a desirable wave-guide material because it possesses excellent electrooptic and acoustooptic properties. In some IO applications, nonplanar structures on $LiNbO_3$ are preferred, but they cannot be fabricated by deposition methods, since such methods do not produce single crystals. Mechanical grinding and ion milling have their shortcomings. This problem was recently solved by the technique of plasma etching [58]. Plasma etching has the advantage of etching selectively, that is, it does not attack masks. Compared with conventional chemical etching, it has the advantage of defining sharp edges.

Employing a CF_4 plasma at 0.3 to 0.55 torr and r.f. power at 200–300 W, plasma etching was reported [58] to be successful on $LiNbO_3$. Ring structures as thin as 3 μm have been fabricated. The etched surface displayed optical quality. Also, a horn-structure ridge-channel wave guide of 20 μm width has been etched and has demonstrated good quality in guiding optical waves.

10. Low-Temperature Ashing

Excited-gas plasma represents a powerful means to induce chemical reactions, treat materials, or provide active species at temperatures much lower than those required when using conventional techniques. By making use of the active species present in a gas plasma, one can carry out low-temperature oxidations or "ashing" of organic substrates. Such reactions usually proceed in a gentle fashion and thus avoid any disturbances of the system such as sintering of fine particles or disturbance of the structure of the developed ash. The decomposition of nongaseous materials in a controlled environment is also an important procedure in the field of spectrochemical analysis. It is often necessary, for example, to separate and recover for subsequent analysis trace quantities of inorganic material present in food and biological and pharmaceutical samples. For an accurate analysis, the procedure must be performed without introducing contaminant substances or inadvertent loss of material.

Investigations of the mineral constituents of biologic specimens generally require removal of the organic substrate. To this end a variety of ashing techniques are presently employed. Each represents a compromise in the complete removal of carbonaceous materials, quantitative recovery of mineral components, retention of mineral microstructure, and absence of chemical contamination. All of the conventional techniques for ashing are based on applying heat directly to the sample. Most of the undesirable side effects could be eliminated by maintaining the sample at low temperatures.

To this goal several patents were granted [59, 60, 61] that employ the use of plasma techniques for low-temperature ashing. This technique has been employed in biological research programs involving samples, such as muscle tissue, liver, fecal matter, whole mice, plant stems, fruit, and milk powder. Substances that are difficult to oxidize, such as carbonized residues, lignin, charcoal, and pyrolytic graphite, have been quantitatively mineralized. The assay of selenium in vegetable matter presents a particularly important example because of the high volatility of many selenium compounds. By the use of O_2 plasma it was possible to determine accurately a few parts of selenium per million parts of alfalfa.

Another example is the oxygen-plasma ashing of human blood to which known quantities of radioactive tracers were added. It permitted the analysis for all metals and recovery of radioactive tracers from blood [62].

Particulate air pollutants collected on filter papers can be dry ashed at 500°C without serious loss of trace metals by volatilization. Comparative studies of particulate matter collected on filter papers in New York City indicate that the results obtained by dry ashing compare favorably with those produced by other accepted methods of analysis [63]. Good recoveries of Pb, Cu, Zn, and Cd from New York City samples were obtained by dry ashing at 500°C for 1 hr.

Other investigators [64] collected weekly samples at three stations in New York City for dry ashing. Metals analyzed included Pb, V, Cd, Cr, Cu, Mn, Ni, and Zn, with detection limits of fractions of $\mu g/m^3$.

The technique of plasma ashing or dry ashing has been applied to a large number of fields, including

Biological: Red blood cells, wine, blood plasma, bile, bones, demineralized bone, dried and ground brain, human hair, human lung, egg albumin, skin, liver, kidney, spleen, heart, testicle, uterus.

Polymers: Virtually all existing polymers have been subjected to dry ashing during the last decade.

Plant materials: Pine needles, alfalfa, casein, rye, book paper, tobacco, tea, soybean, corn oil, plankton, clover, rubber leaf, and so forth.

Geological and mineral products: Dirty crankcase oil, coal, graphite, floral coke, carbon black, oil shale, diamond chips, potting soil, cobalt and nitrogen-containing organic compounds, and so forth

Biochemical: Powdered milk, lactose, ice cream mixes.

Other areas of plasma ashing include microincineration of biological specimens to produce ash patterns that show in mineral localization in situ. This technique has also been employed to retain a three-dimensional ash structure. For example, the structure of a polyurethane foam was fully retained after plasma ashing [65]. The microcellular membranes of the foam were ashed away by the glow discharge, while the slight amount of mineral material contained in the membranes formed into connecting microfilaments and strands to retain the original shape of the foam specimen.

Similarly, Thomas [65] was able to preserve the ash structure in three dimensions on rice hulls chiefly because of the high content of bioincorporated silica. On a smaller scale, plasma incineration was able to preserve the three-dimensional ash structure of bacterial spores. Another interesting experiment on microscopic aquatic animals displayed the delicate three-dimensional ash skeleton of a rotifer jaw.

11. Textile Treatment

Modification of textile fibers by glow-discharge grafting has attracted considerable interest in more recent times. Thus, it has been observed that a fraction of 1% of acrylic acid plasma-grafted onto polyester fabrics renders the fiber as wettable as cotton [66].

Other synthetic fabrics, such as nylon and orlon, have also been altered by this free-radical grafting process, thus providing new marketable improvements in various properties, such as moisture retention and reduced static charge. Cotton itself, it is said [66], can be converted into a "superwashable" fabric with all its original bulk and texture retained.

Uniform grafting of polymerized acrylic acid onto polymer surfaces from polyolefins, polyester, Teflon, and other substances has been found to impart properties, such as wettability, adhesiveness, dyeability, and ease of print-

Table 8.5. Soil release as a function of graft level in polyester tricot

Acrylic acid add-on % W/W	Soil release rating AATCC test 130 (1969)	
0.14	4.5	remaining traces barely detectable
0.073	4.0	faint outline of stain
0.032	2.5	moderate (similar to cotton)
0.015	0.5	no significant improvement
Untreated	0	no release of stain

Source: A. Bradley and J. D. Fales, *Chem. Tech.,* p. 237 (April 1971).
Reprinted with permission. Copyright 1971 American Chemical Society.

ing. The bonds of such grafted structures have been shown [66] to survive over 30 washings without losing more than 10–25% of the original weight gain.

Another important factor in such textile grafting is the quality of "stain release." Thus, Bradley and Fales [66] reported that about 0.04% of "add on" by plasma grafting sufficed to obtain complete removal of dirty motor oil stain in double-knit polyester fabrics. Polyester tricot released oily stains in degrees depending on the graft level (table 8.5). Untreated cotton broadcloth gives a soil release rating of about 2, which can be improved to a rating of about 4 by a light acid graft.

A glow-discharge grafting machine for treating single rolls of fabric up to 200 yds long and 66 in. wide has been in operation for nearly 10 years [67]. It consists essentially of an electrode and roller assembly that can be wheeled in and out of a large steel tank. The tank can be evacuated and filled with monomer vapor and inert gas at reduced pressure. Thirty-two electrodes are energized in parallel lines via insulated cables. The luminous plasma can be regulated while the fabric is moving.

12. Plasma Detoxification and Air Purification

Recently, Wydeven and Flamm [68] developed a plasma reaction system that can remove trace contaminants from spacecraft, submarines, and other closed environments by oxidizing the contaminants to produce carbon dioxide and water.

The contaminants range from metabolic products, such as alcohols, esters, hydrogen sulfide, and ammonia to solvents, such as Freon, aromatics, alcohols, and ketones. The oxidized contaminant products are removed from the reactor by a scrubber. The system consists of a reactor powered by a variable r.f. source. A helical resonator generates the high voltage to produce a glow discharge in the reaction chamber. The resonator is connected to the r.f. source through a directional wattmeter.

Table 8.6. Pesticides and hazardous wastes for detoxification tests

Classification	Material	Form Tested
Organophosphorus pesticide	malathion (95% min. purity)	pure liquid
Chlorinated hydrocarbon waste	PCBs (polychlorinated biphenyls)	liquid mixture
Brominated hydrocarbon rodenticide	methyl bromide (99.5% min. purity)	commercial gas
Heavy metal fungicide	phenylmercuric acetate (30% PMA solids)	commercial aqueous
Chlorinated hydrocarbon pesticide	kepone (80% active ingredient, 20% clay)	1. commercial powder 2. laboratory aqueous dispersion 3. laboratory methanol solution
Polyaromatic red dye mixture	55.4% xylene azo-β-naphthol 18.9% 1-methylaminanthraquinone 18.0% sucrose 1.8% graphite 5.9% silica binder (KClO$_3$ oxidant excluded)	1. laboratory aqueous dispersion 2. laboratory methyl ethyl ketone solution

Source: L. J. Bailin, B. L. Hertzler and D. A. Oberacker, *Envir. Sci. and Technol., 12*(6), 673 (1978). Reprinted with permission. Copyright 1978 American Chemical Society.

Table 8.7. Additional plasma processes

Feed	Product	Some uses
$TiCl_4$ + Na	pure titanium	aerospace, chemical industry
Coal/oil	carbon black	pigment, tire filler
Tin slag	tin oxide	high-grade tin ore
Iron ore + H_2 Iron ore + rich flame }	iron	steel production
$ZrCl_3$ + Na	pure zirconium	nuclear reactors
MoS_2	molybdenum	steel additive
Calcium fluorophosphate	{ calcium fluoride { phosphorus	steel mfg., aluminum smelting fertilizer
Titanium, tungsten, tantalum + carbon or nitrogen	metallic carbides and nitrides	tool steel additives

Source: J. Lawton, *Phys. in Technol.,* p. 190 (Sept. 1975). Copyright © 1975 by The Institute of Physics.

As the oxygen atmosphere containing contaminants enters the reactor, the resulting glow discharge causes oxidation of the contaminants to produce water and CO_2. In tests on prototype models, the concentration of methane was reduced by three orders of magnitude, with the discharge operating at 50–300 torr pressure. The removal rate is reported [69] to be independent of the contaminant concentration, offering an important advantage over other air purification systems.

A microwave-induced oxygen plasma was recently applied with success to detoxification of pesticides and hazardous wastes [69]. The contaminants were passed through a reactor that permitted a throughput of 1–7 lbs/hr. The materials that were detoxified or decomposed in the plasma chamber are listed in table 8.6 [69].

When oxygen is used in the plasma, its atomic species causes the rapid oxidation of organic compounds and, with the assistance of the energetic free electrons, final degradation into CO, CO_2, and H_2O. Malathion yielded, in addition to the aforementioned end products, SO_2 and metaphosphoric acid, $PCB-O_2$ converted into Cl_2, HCl, CO_2, CO, and H_2O. For phenylmercuric acetate-O_2 the formation of metallic Hg may be due to HgO as an intermediate, followed by its decomposition and dissociation between 440–610°C.

Other promising plasma chemical processes are listed in table 8.7.

References

1. National Materials Advisory Board. *Natl. Res. Council Publ. NMAB–349;* Natl. Res. Council: Washington, D.C. (1979).

2. C. K. Jones, D. J. Sturgess, and S. W. Martin. *Metal Progress, 8,* 186 (1972).

3. F. E. Gifford. *J. Vac. Sci., Technol., 10,* 85 (1973).

4. B. B. Buecken, *Die Technik, 33*(7), 395 (1978).

5. G. Dernaley, *Trans. Inst. Metal Finishing, 56,* 25 (1978).

6. J. K. Hirvonen, *J. Vac. Sci., Technol., 15,* 1662 (1978).

7. Wen-Wei Hu, C. R. Clayton, H. Herman, and J. K. Hirvonen. *Scripta Met., 12,* 697 (1978).

8. N. E. W. Hartley. Presented in part at the Conference on the Application of Ion Beams to Materials, Coventry, England (Sept. 1975).

9. P. D. Townsend. *J. Phys. E. Sci. Instr., 10,* 197 (1977).

10. G. F. Bason. *Trans. of Asst., 14,* 133 (1928).

11. A. M. Taylor. Advanced Vacuum Systems, Ayer, Mass. (1980).

12. C. Braganza, S. Veprek, E. Wirz, H. Stüssi, and M. Textor. Presented in part at the 4th International Symposium on Plasma Chemistry, Zurich, Switzerland (Sept. 1979).

13. B. Edenhofer, *Metal Progr.,* p. 181 (Mar. 1976).

14. K. S. Cho and C. O. Lee. Trans., *ASME, 102,* 222 (1979).

15. T. Sone and K. Yamanaka. *J. Japan Inst. Met., 41,* 620 (1977).

16. K. Imai, *J. Japan Inst. Met., 7,* 137 (1968).

17. W. L. Grube and J. G. Gay. *Metallurg. Trans. A, 9A,* 1421 (1978).

18. F. E. Gifford and W. Cooley. *J. Vac. Sci. Tech., 11*(2), 511 (1974).

19. R. A. Kant, J. K. Hirvonen, A. R. Knudson, and J. S. Wollam. *Thin Solid Films, 63*(1), 27 (1979).

20. LFE Corp. Bull. #8200–SB1 (April 1976).

21. LFE Corp. Bull. #8214–TA1 (June 1975).

22. A. Bradley. *Ind. Eng. Chem. Prod. Res. and Dev., 9,* 101 (1970).

23. L. Kronick and M. E. Schafer. *J. Appl. Polym. Sci., 13,* 249 (1969).

24. L. Liepins and J. Kearney. *J. Appl. Polym. Sci., 15,* 1307 (1971).

25. L. Liepins and H. Yasuda. *J. Appl. Polym. Sci., 15,* 2957 (1971).

26. See chapter 7, section 7.

27. J. R. Hollahan, B. B. Stafford, R. D. Fall, and S. T. Payne. *J. Appl. Polym. Sci., 13,* 807 (1969).

28. F. Y. Chang, M. Chen, and A. T. Bell. *J. Appl. Polym. Sci., 17,* 2915 (1973).

29. J. R. Hollahan, T. Wydeven, and C. C. Johnson. *Appl. Opt., 13,* 1844 (1974).

30. P. K. Tien. *Sci. Amer., 230*(4) (April 1974).

31. P. K. Tien, G. Smolinski, and R. J. Martin. *Appl. Opt., 11,* 637 (1972).

32. P. K. Tien, R. J. Martin, and G. Smolinski. *Appl. Opt., 12,* 1909 (1973).

33. P. K. Tien, S. Riva-Sanseverino, R. J. Martin, and G. Smolinski. *Appl. Phys. Letters, 24,* 547 (1974).

34. R. H. Hansen, J. V. Pascale, T. DeBenedictis, and P. M. Rentzepis. *J. Polym. Sci., A-3,* 2205 (1965).

35. See chapter 7, section 2B.

36. C. Y. Kim. *J. Appl. Polym. Sci., 15,* 1365 (1971).

37. E. L. Lawton. *J. Appl. Polym. Sci., 18,* 1557 (1974).

38. D. F. O'Kane and K. L. Mittal. *J. Vac. Sci., Technol., 11*(3), 567 (1974).

39. See chapter 7, section 1.

40. A. Erb. *J. Phys. Chem., 69,* 1306 (1965).

41. T. M. Fowkes. ASTM Spec. Tech. Publ. No. 360, p. 20 (1963).

42. M. K. Bernett and W. A. Zisman. *J. Phys., Chem., 74,* 2309 (1970).

43. R. S. Thomas. In *Techniques and Applications of Plasma Chemistry,* ed. J. R. Hollahan and A. T. Bell. Wiley: New York (1974).

44. H. Orth. *Z. Wiss. Mikr. mikr. Tech., 70,* 179 (1970).

45. A. Ye Chalykh. *Vosokomol. Soyed., A16,* 1289 (1974).

46. E. W. Fischer, H. Goodar, and G. F. Schmidt. *Kolloid Z., Polymere, 226,* 30 (1968).

47. L. I. Bazruk, N. Okui, and T. Kawai. *J. Macromol. Sci., Phys., B10, 543* (1974).

48. S. B. Warner, D. R. Ullmann, and L. H. Peebles. *J. Mater. Sci., 10,* 758 (1975).

49. M. R. Padhye, N. V. Bhat, and P. K. Mittal. *Text. Res. J., 46,* 502 (1976).

50. A. Garton, P. Z. Sturgeon, D. J. Carlsson, and D. M. Wiles. *J. Mater. Sci., 13,* 2205 (1978).

51. R. L. Shepard. U.S. Pat. 3,415,986 (1968).

52. M. R. Padhye, M. U. Bhat, and P. K. Mittal. *Text. Res. J., 46,* 502 (1976).

53. P. R. Blakey and M. O. Alfy. *J. Text. Inst.,* No. 1, 38 (1978).

54. R. C. Booth and C. J. Heslop. *Thin Solid Films, 65,* 111 (1980).

55. C. M. McKenna and H. K. Willcox. U.S. Pat. 4,183,780 (1980).

56. R. E. Hampy. Sandia Labs., Res. Rept. No. SA–ND 77-2070 (1978).

57. R. W. Kirk. In *Techniques and Applications of Plasma Chemistry,* ed. J. R. Hollahan and A. T. Bell. Wiley: New York (1974).

58. C. E. Lee and C. L. Lu. *Appl. Phys. Lett., 35*(10), 756 (1979).

59. J. R. Hollahan. U.S. Pat. 3,428,548 (1969).

60. G. J. Gorin. U.S. Pat. 3,619,403 (1971).

61. C. E. Gleit. U.S. Pat. 3,547,802 (1970).

62. C. E. Gleit. *Amer. J. Med. Electronics,* p. 112 (April 1963).

63. T. Y. Kometani, J. L. Bove, B. Nathanson, S. Siebenberg, and M. Magyar. *J. Air Poll. Contr. Ass., 6*(7), 617 (1972).

64. T. J. Kneip, M. Eisenbud, C. D. Strehlow, and P. C. Freudenthal. *J. Air Poll. Contr. Assn., 20*(3), 144 (1970).

65. R. S. Thomas. In *Techniques and Applications of Plasma Chemistry,* ed. J. R. Hollahan and A. T. Bell. Wiley: New York (1974).

66. A. Bradley and J. D. Fales. *Chem. Tech.,* p. 221 (April 1971).

67. Surface Activation Corp., 1150 Shames Dr., West Broadway, New York 11590.

68. T. J. Wydeven, Jr. Stanford Univ., Calif.; D. L. Flamm, Ames Res. Center, NASA.

69. L. J. Bailin, B. L. Hertzler, and D. A. Oberacker. *Envir. Sci. and Technol., 12*(6), 673 (1978).

Index